# CORAL LIVES

# CORAL LIVES

## Literature, Labor, and the Making of America

MICHELE CURRIE NAVAKAS

PRINCETON UNIVERSITY PRESS
PRINCETON & OXFORD

Published by Princeton University Press
41 William Street, Princeton, New Jersey 08540
99 Banbury Road, Oxford OX2 6JX

press.princeton.edu

ISBN 9780691240114
ISBN (pbk.) 9780691240091
ISBN (e-book) 9780691240107

British Library Cataloging-in-Publication Data is available

Editorial: Anne Savarese and James Collier
Production Editorial: Natalie Baan
Text and Cover Design: Heather Hansen
Production: Erin Suydam
Publicity: William Pagdatoon
Copyeditor: Ellen Hurst

Jacket image from Louis Ferdinand comte de Marsilli, *Histoire Physique de la Mer*, 1725. Courtesy of Bibliothèque nationale de France.

Title page image: *Memoirs of the Museum of Comparative Zoölogy, at Harvard College, Cambridge, Mass, Vol. 7, No. 1* (1880). Pl. XIX, No. 9. Ornament image (p. 13): Louis Ferdinand comte de Marsilli, *Histoire Physique de la Mer*, 1725. Tab. XXVIII, Fig. 129, No. 4.

This book has been composed in Adobe Text Pro with Sirenne Six MVB

Printed on acid-free paper. ∞

Printed in the United States of America

10 9 8 7 6 5 4 3 2

*For Gene*

So wonderful thou art, that nought beside
Seems worth a song in the wide rolling sea.

—"The Coral Worm" (1836) by Mrs. Thompson
of Tioga County, New York

# CONTENTS

# Illustrations

Color plates for figures 1.1, 1.2, 1.5, 1.10, 1.11, 1.20, 1.24, 1.25, 5.2, and 5.5 follow page 98.

# Coral Lives

# Introduction

In the twenty-first century, people in many parts of the world see coral only on rare occasions. A visit to the aquarium or natural history museum brings coral reefs or coral specimens close. A snorkeling or scuba excursion briefly reminds us of the shocking vibrancy of the underwater world. An alarming environmentalist documentary or newspaper report confirms that coral is dying, as warming seas bleach the world's reefs beyond recovery.

Yet in nineteenth-century Europe and North America, coral was every-where. Women and girls—wealthy, working-class, and enslaved—wore coral necklaces, pins, rings, earrings, and bracelets, jewelry far more accessible and affordable than gold or gemstones. A demand for these and other coral objects drove the global coral trade, then centered in Italy, where Mediterranean coral fishers brought yearly harvests of raw coral to "coral workshops" for workers to cut and polish and then prepare for packing aboard ships destined for foreign ports and markets. Reef ecosystems flourished throughout the Mediterranean Sea and in the warm, shallow waters of the Indian Ocean, the Pacific Ocean, and the Caribbean Sea, all of which were major sources of coral reef specimens retrieved from the seafloor by local divers to supply the curiosity cabinets of naturalists and tourists. Coral served as currency in the transatlantic slave trade; between the fifteenth and eighteenth centuries Europeans exchanged coral for persons, capitalizing on the value of coral beads within certain cultures of Africa, such as in the Kingdom of Benin in what is presently southern Nigeria. Museum visitors peered at display cases of coral extracted from the Pacific by scientific expeditions. People were eager to learn about coral's natural history of defying taxonomic boundaries separating animal, vegetable, and mineral from the classical period through the Enlightenment. And children in wealthier families cut their first teeth on the "coral and bells," a combination teething aid, toy, and talisman that evoked the classical myth of coral's "birth" from Medusa's blood and was believed to ward off maladies both physical and spiritual.

On the whole Europeans and North Americans not only saw and touched coral more often than we do today, but they also had good reasons to think more about the nature and growth of living, reef-building corals in particular. As a well-known cause of shipwrecks, reefs raised the question of where and how coral grows, a matter of economic concern to traders, of terror to navigators, and of sovereignty to any empire with oceanic ambitions. These and other reasons drove intense global interest in scientific theories of coral reef formation, such as those famously advanced by Charles Darwin and Charles Lyell during the 1830s and 1840s. And these theories, in turn, engaged a wide public during a time when the sciences were not yet the specialized domain of experts, but rather, as literary scholar Laura Dassow Walls explains, "part of the buzz and flux of the newspapers, parlors, and periodicals, right alongside— often the subject of—poems and stories and gossipy fillers."[1]

Certainly the average person did not have a detailed understanding of coral biology.[2] Yet it was common knowledge that tiny polyps, discovered during the eighteenth century, somehow produce massive and ever-expanding structures by continually fusing together, while collectively making a reef from their bodies, over a timespan so vast that humans could scarcely fathom it. This process was popularly imagined as "labor" or "work." Descriptions of polyps toiling away, "down, down so deep" in the sea—to quote the lyrics of one song especially beloved by generations of US schoolchildren—filled countless poems, short stories, novels, periodical essays, and other widely circulating media.[3]

For many reasons, then, people once encountered coral more frequently and knew and imagined much more about its nature, meanings, histories, and uses than most of us do today. These conditions set the stage for the particular phenomenon that this book explores: in the nineteenth century a powerful set of ideas about coral shaped US thinking and writing about politics, broadly defined as a system of managing and distributing finite resources and care. Thus, while this book tells the story of coral as at once a global commodity, a personal ornament, an essential element of the marine ecosystem, and a powerful political metaphor, it also asks us to consider what we of the Anthropocene might learn from the forgotten human lessons once encoded in coral, even as coral itself vanishes.

One of the most popular political analogies that coral inspired during the nineteenth century involved the comparison of human society to a coral reef. The analogy usually suggested the power of collective labor for common good. In a chapter on cooperative labor in *Capital* (1867), for example, Karl Marx borrows this analogy by comparing humans to reef-making polyps: to make the point that collective labor promotes collective thriving, Marx cites a contemporary textbook on political economy that describes "mighty coral reefs rising" from the work of polyps who, though individually "weak," are strong in the aggregate.[4] Marx's point is that coral offers humans a better model of labor and politics than capitalism, for the labors of coral serve not the "mastership of one capitalist," but rather the sustenance of the whole, the work of each individual polyp ultimately enriching all.[5] Variations on that political romance of the coral collective abound in nineteenth-century print culture on both sides of the Atlantic.

Yet in an enormous range of nineteenth-century US reflections on coral reefs, that bright romance collides with and founders on a darker vision of life-consuming labor performed by the many for the benefit of the few. In one of the most popular nineteenth-century US poems about coral, for example, reefs exhibit a system of production in which collective labor *appears* to yield collective thriving, while in reality it requires one group of workers to give their lives and labors in full to a robust and expansive foundation that they can never enter alive. An instant transatlantic success upon its 1826 publication in a Connecticut newspaper, Lydia Huntley Sigourney's "The Coral Insect"—a popular term for the coral polyp—describes a "race" of beings that "toil" to "build" the massive reef. They do so collectively, ultimately producing a "vast" work of lasting "wonder and pride."[6] Thus far, just as in *Capital*, reefs are nature's celebration of communal labor for the common good.

In this widely reprinted poem, however, as in so many other nineteenth-century US reflections on coral, the details of reef formation disclose another story. Coral insect work is ceaseless and unvarying: the poem opens with repetitive labor—"Toil on! toil on!"—and closes with its continuance in the rhythmic "Ye build,—ye build." That perpetual toil goes unacknowledged, absent from human sight and memory: it is "secret," "noteless," and "unmark'd." Generation after generation, from birth until death without leaving, the workers build a structure that excludes them, for they "enter not in" but rather "fade" into the "desolate main," where they "die."[7] Meanwhile, the reef rises from their laboring bodies, which endlessly merge to become a coral island that supports those who did not produce it and do not remember who did.

Sigourney's poem, unlike Marx's treatise, does not directly compare polyps to persons or reefs to human polities. Instead, the poem explores a key conceptual question raised by coral's natural history: How can countless small and finite beings create and sustain a single, massive, enduring, and growing structure? The poem was published in a historical moment when this question was of the most pressing political and cultural relevance to a polity dedicated to collective thriving and sustained by the labors of millions who could never fully belong to the structure that they built. It answers the question by imagining extractive labor as a necessary condition of the most robust foundations. And it shows just how easily such labor may exist alongside, become obscured by, and even promote celebrations of communal labor for common good.

Sigourney's poem is far from singular in that regard. Rather, it belongs to a vibrant tradition of nineteenth-century US accounts of coral reefs in which the vision of a cooperative coral collective can be sustained as long as one does not look beneath the waves to perceive that a reef can emerge into a robust island by continually and silently extracting the labors and bodies of countless millions of workers. By rehearsing that account of coral in innumerable texts in multiple genres across the long nineteenth century, Americans repeatedly described an extractive labor relation that strikingly resembled the chattel slavery that was then sustaining and expanding the US.[8] Moreover, through repetition, that labor relation became familiar, and even routine.

Sigourney's poem vividly registers just one of several branches of US political thinking about coral that emerged in the polity's written and visual culture across the long nineteenth century. During this period, coral's "cultural biography"—briefly defined as the various histories and uses of coral that generate its social meanings—intersected with political pressures and debates specific to a culture formally dedicated to common good yet increasingly indebted to "slavery's capitalism," the unprecedented, entwined expansion of slavery and industrial capitalism between the 1790s and 1860s.[9] Tracking these intersections across US writing and visual culture, accounting for why they recur, and explaining their political significance among different groups of Americans at different moments in the long nineteenth century is the central work of this book. For even, perhaps especially, when coral appears as mere ornament, curiosity, or romantic metaphor, it almost always also reflects and shapes complex conceptual thinking about labor and life, individual and collective, alternately generating visions of the common good *and* numerous forms of reckoning with or refusing characteristically US capitalist coercions and exclusions—sometimes within the very same text.

We already know that nineteenth-century Americans vigorously debated the benefits of industrial capitalism, broadly defined as a system of economic behaviors and relations that arose and expanded globally between 1790 and 1860 by mobilizing, and then rationalizing, very different kinds of unfree labor, among which slavery was by far the most exploitative, violent, and formative. We know that Americans continued to debate these behaviors and relations as they spread beyond the plantation and even beyond the legal end of slavery. Relatedly, scholars have documented significant popular contests over one of the most pernicious ideologies that arose to rationalize these coercive relations: biological essentialism, a strictly biological account of human life that ranks some persons as more naturally suited to bodily labor and others to a life of the mind.[10] We know, in other words, that the US is founded in unfree labor relations of very different kinds—including slavery, wage labor, and the unpaid reproductive and domestic work of women—and that many struggled to resist or rationalize such labor through various scientific, philosophical, and social discourses because coercive labor is at variance with the vaunted ideal of common good.[11]

Coral, however, allows us to see how such a country and some of its most powerful and pervasive (yet elusive) colonial logics of race, gender, and class were alternately conceived, forged, circulated, normalized, rationalized, contested, and refused—not only in the period's more obviously political writings but also in a broad range of seemingly ephemeral musings on an everyday material that frequently appears to have nothing to do with humans at all.[12] As writers of various identities, political affiliations, intentions, and intended audiences considered coral, coral's material features and histories in turn shaped their thinking, producing new accounts of coral that are also infused with claims of striking political relevance. In some texts, that relevance is more apparent than in others because many US writers directly compare coral polyps to humans, and often to specific groups of people working to sustain and expand the nineteenth-century US. In other texts, that relevance emerges as an unmistakable continuity between the imagined labor conditions of coral reefs and the actual labor conditions in US society. In still other texts that relevance becomes visible only when we learn a specific history of coral's use and meaning—a particular part of coral's cultural biography—that was familiar to many nineteenth-century Americans yet is largely forgotten today.

The political claims that emerge from manifold intersections of coral and US politics are familiar, unsettling, and surprising by turns. In some cases, coral abetted well-documented forms of colonial thinking by lending new force to various logics of race, gender, and class that rationalized the polity's ongoing dispossessions and exclusions. In other cases, coral generated challenges to these logics from unlikely quarters, since many who benefited socially and economically from the combined rise of slavery and industrial capitalism also questioned these interlinked systems through coral. And more hopefully for our present world, coral frequently prompted writers, artists, and activists to imagine more just political arrangements and social possibilities that remain to be put widely into practice.[13]

Whatever form the country's political thinking with coral took, our attention to it reveals that a broad cross section of the population, across boundaries of race, gender, and class—poorer and richer, worker and employer, union leader and lawmaker—repeatedly and tacitly confronted a reality that was too rarely stated explicitly: a country formally dedicated to unprecedented liberty required marginalized persons to labor in the service of a polity that refused to incorporate them politically and socially. That acknowledgement was once a ubiquitous, widely shared, familiar part of everyday life in the US, as close as a coral specimen or a song about coral reefs. And it speaks to a complex problem that persists at the center of US life: our longing for the collective runs deep and too easily elides the coercions sustaining it.

## METHODS AND APPROACHES TO REEF THINKING

How did coral, of all materials, come to quietly carry complex political imaginings of labor and life, individual and collective—even when coral appears to be merely ornamental? Which particular groups of humans and forms of labor did Americans tend to associate with coral, and why? How can we recover the varieties of political imagining once encoded in coral as a result of coral's manifold histories? And why do those past imaginings matter now? While it will take an entire book to answer these questions in full, they require a preliminary explanation of my methods for uncovering the polity's political thinking with coral.

This book began in earnest when I began to take a biographical approach to coral, seeking to understand the social meanings of coral in the nineteenth-century US, as those meanings emerged and changed across time and location, among different groups of people, and in response to various sources of knowledge about coral—whether scientific, literary, economic, cultural, or otherwise—in tandem with contemporary social and political exigencies.[14] This approach has required closely reading texts about coral by numerous nineteenth-century US writers, artists, and activists, and then trying to follow wherever their representations of coral pointed.[15]

One part of that project has involved becoming conversant with many sources of knowledge about coral that were more familiar to nineteenth-century Americans than they are to most people living in the US today. These sources, which I discuss at length in chapter 1, include the natural history of coral from the classical period through the Enlightenment; Enlightenment debates about coral taxonomy and coral polyps among prominent European naturalists from Trembley to Linnaeus; nineteenth-century scientific studies of reef formation by Darwin and others; British poetry about Pacific coral reefs; British and American travel writing about Pacific coral reefs and islands by scientists and explorers; the European medicinal and spiritual practice of wearing red coral to ward off the "evil eye" and protect the bodies of mothers and children; the industry of Mediterranean coral fishing that supplied red coral to the global coral trade before 1900; oral traditions and ritual practices surrounding red coral within specific cultures of early modern Africa, and later within particular communities in the Black Atlantic diaspora from Jamaica to New Orleans; the role of coral as currency in the transatlantic slave trade from the fifteenth to the eighteenth century; and the practice of coral diving among free and enslaved populations in the Caribbean to supply coral specimens to tourists and naturalists. All these practices, uses, and meanings produced knowledge about coral that informed nineteenth-century American conceptions and representations of it.

But another part of that project has involved realizing that the cultural biography of coral is inseparable from colonial violence at almost every turn—and then using that realization to uncover some particular conceptual and material relations between coral and colonial violence and letting them guide my interpretations of imaginative reflections on coral. While no domain of human experience or knowledge is fully legible apart from the history and ongoing legacies of colonial violence—as scholars working in Black and Indigenous studies have established—there are several specific reasons that Western knowledge about coral is especially imbricated with that history.[16]

First, as is the case with many types of natural specimens that European natural historians turned into their objects of study, local Black and Indigenous persons, both free and enslaved, frequently supplied coral specimens, alongside useful knowledge about coral itself, to naturalists working in the Caribbean and the Pacific.[17] These local populations also supplied decorative coral specimens to US tourists, who would then display them on mantles, bookshelves, and curiosity cabinets in their homes and classrooms and frequently use them to teach or speculate about the nature of coral. Nineteenth-century knowledge of coral, both scientific and popular, was thus quite literally indebted to the skills, knowledge, and labor of Black and Indigenous persons.

Second, the practice of eighteenth-century natural history provided a ready logic to nineteenth-century racial scientists, while the particular Enlightenment debate about coral taxonomy shared specific conceptual terrain with nineteenth-century debates about human taxonomy. Biological essentialism, a biological account of human life that nineteenth-century American racial scientists embraced and developed under pressure to rationalize the expansion of slavery and other colonialist projects, borrowed heavily from systems of taxonomic classification proposed by earlier generations of natural historians such as Linnaeus.[18] The enormous epistemological shift that attended the nineteenth-century emergence of the life sciences, and involved the transformation of earlier taxonomic systems into rigid tools for hierarchizing humans, has been exhaustively documented.[19] I raise it here, first, to remind us that American reflections on natural history during the period under consideration were never uninformed by the emergent scientific racism that rationalized slavery, and second, to introduce a more specific relation between racial taxonomy and coral taxonomy that developed during the period. For biological essentialism turned on the claims that race and gender are visible, and that they are visible indices of what is harder to see—qualities such as character, intelligence, and creative capacities—claims destabilized, according to many US writers, by the natural history of coral as an animal that even the luminaries of Enlightenment science had repeatedly misidentified as a stone or a plant based on how it looks and feels. Within a culture that increasingly ranked and valued persons based on the color, shape, size, and texture of their physical features, reflections on coral's natural history were rarely about coral alone but also bore directly on current debates about slavery and some of its rationalizing logics and ways of seeing and evaluating.[20]

Third, transatlantic slavery produced divergent knowledges and practices surrounding coral, while partly sponsoring Mediterranean coral fisheries that supplied most of the world's coral objects before 1900. Indigenous African value systems rendered red coral beads an important currency in the transatlantic slave trade between the fifteenth and eighteenth centuries. These systems made it possible for European traders to purchase humans with red coral, much of it retrieved by Mediterranean coral fishers hired by powerful European mercantile companies that facilitated the slave trade, including Great Britain's Royal African Company. These same Indigenous value systems also persisted beyond the exchange of coral for persons, however, shaping the styles of dress among Black women throughout the Black Atlantic diaspora who wore red coral well into the nineteenth century to preserve a link to the past and affirm their multiply informed racial, social, and cultural diasporic identities. Thus, in the nineteenth-century US, coral alternately signified the reduction of Black persons to commodities and the refusal of the logic of reduction, depending on whom one asked.

These are some of the major reasons that we must read nineteenth-century US reflections on coral in relation to histories of colonial violence. These relations help explain why so many different groups of people during this period considered coral strikingly relevant to questions of human labor. These people include white women, such as poet Lydia Huntley Sigourney, for whom coral sustained the idea that the most robust structures emerge by consuming and excluding countless "laborers" over many generations. They include Black women, such as poet and activist Frances Ellen Watkins Harper, who borrowed the reef analogy to point out the exclusionary social and political relations that subordinated Black persons long after the legal end of slavery.

They include the Coral Builders, a Black literary society in late nineteenth-century Cleveland that took up the metaphor of coral to imagine and enact a more just distribution of essential labor. And they include white abolitionist Harriet Beecher Stowe and Black political activist Charles W. Chesnutt, writers for whom red coral beads generated the fleeting but formative recognition that Black life is also lived beyond the reality and conceptual framework of slavery.[21]

Yet finding the complex reflections on human politics encoded in disparate reflections on coral requires more than close reading. It also requires key insights from Black studies, and from studies of material culture and race in North America and the broader Atlantic world.

A central goal of Black studies, as Alexander G. Weheliye reminds us, drawing on the work of Sylvia Wynter, is to track the conceptual frameworks, logics, and processes that produce human hierarchies by stratifying humanity into "full humans, not-quite-humans, and nonhumans"—and by doing so in subtle ways.[22] Those hierarchies can emerge, circulate, and gain credibility—and alternately be contested, refigured, and refused—through repetition across time, not only in overtly political texts but also in everyday reflections on topics that are apparently apolitical. Powerful modes of ranking and devaluation, of dividing and hierarchizing, Lindon Barrett suggests, often transpire in the most quotidian ways.[23] In the nineteenth-century US, racist presumptions about persons of African descent in particular were never confined only to texts that deal explicitly with the question of race.[24] A major premise of this book is that human differences are sometimes intensely produced and reinforced, challenged and refused, through texts that may not appear to be about race, or even about humans at all.[25]

Black studies scholars also remind us, however, that the human hierarchies forged under slavery and scientific racism have never wholly defined the lives of Black persons. As Katherine McKittrick observes, both biological determinism *and* the scholarly critique of biological determinism risk violently diminishing Black persons by suggesting that the history and afterlives of slavery can fully explain and contain Blackness.[26] This book tries to heed McKittrick's warning by reading representations of coral in relation not only to violent colonial practices and frameworks but also to Black self-representation and community formation within the US, to meanings of coral developed by Black women across the Black Atlantic diaspora, and to fleeting moments of white recognition that Black life far exceeds the US colonial logic of race as biology.

Tracking how coral, in particular, alternately abets and unsettles colonialist frameworks within the nineteenth-century US would also be impossible without the methods of material cultural studies, which interprets materials and the metaphors they sponsor in relation to multiple human histories and practices. Of particular importance to this book is cultural historian Robin Bernstein's concept of "scriptive things," items of material culture that encourage or "script" certain ideas and behaviors because of their particular features in combination with their documented uses and meanings among human communities across time and contexts.[27] Put another way, things carry meanings because of how different groups of people have used them over time, and consequently, things may transmit those meanings beyond the intent or understanding of any given actor or author.[28] As a marine material with a global history spanning many centuries and cultures, coral carried many meanings—of nature and culture, individual and collective, labor and life—that took shape far beyond the US polity's cultural, political, temporal, and spatial boundaries. To track potential meanings of Blackness in particular that coral generated beyond those boundaries, I have relied on the methods of scholars of material culture in African and African Diaspora communities in the Atlantic world. Of special importance in that regard is Akinwumi Ogundiran's and Paula Saunders's concept of "Black Atlantic ritual." Ogundiran and Saunders argue that fragments of information about the memories, behaviors, values, and practices of persons forcibly entangled with commerce, commodification, Middle Passage, and slavery are encoded in the materials that mattered to them.[29]

By recovering some varieties of political thinking once encoded in coral in the US, this book cannot by itself solve the problem to which it relentlessly returns, the country's persistent failure to create a truly just collective. After all, even the more optimistic varieties of political imagining with coral that I examine did not translate into wider practice or prevent the enslavement of millions of persons of African descent during a period that is rightly known for forging violent cultural logics of race and gender that persist into our present. Yet the goal of historically oriented humanities scholarship is not necessarily to provide us with ready-made solutions, but rather, as historian Dominick LaCapra reflects, to help us think through problems that "do not lend themselves to a definitive solution or first-time discovery," have been "repeated (and repeatedly thought about) with variations over time," and remain "alive and pressing."[30] This book aids our thinking about collective thriving in the US by using coral to illuminate, clarify, and historically contextualize that topic, to unearth overlooked solutions imagined by past generations, and to encourage us to understand ourselves in relation to, even as inextricable from, that past.[31] For even as this book charts new origins and trajectories of exclusionary thinking and practice in the US, it also reveals that a refusal to categorize, know, and rank life is also a vital part of the US polity's cultural past, and so that refusal must become a "condition of the present."[32]

## SOURCES AND ORGANIZATION

Of the hundreds of widely circulating written and visual reflections on coral and reefs consulted for this study across nearly ten years of research at scores of US (and some European) libraries, museums, and other cultural institutions, I have chosen to focus on those that most vividly convey the most interesting, surprising, and relevant lines of popular thinking about coral, labor, and politics in the US between roughly 1790 and 1900. These sources span multiple media and genres, including natural history, periodical essay, poem, song, short story, decorative object, specimen, painting, Sunday school pamphlet, newspaper article, novel, sermon, reference work, stereograph card, travel account, and trade dictionary. Nearly all circulated widely during the period in question, which means that fewer than I have wished were produced by persons marginalized by race, gender, and class. To be more particular about demography: about half of the sources included were created by white men; roughly a quarter by white women; roughly a quarter by anonymous creators; and a small percentage by Black men and women. Finally, while this study centers largely on US literature and culture between 1790 and 1900, its temporal and geographic span is much broader because so much nineteenth-century thinking on coral, labor, and politics derives from meanings of coral forged before 1790 and beyond the borders of the continent. Thus, I have tried as much as possible to take an oceanic approach to nineteenth-century US writers and artists by tracking multiple associations they made with coral; this approach has meant following their imaginations across the multiple centuries, continents, and cultures that brought coral itself, and ideas about coral, into everyday US life.

Immersing the reader in the global circulations of materials, bodies, and ideas that endowed coral with the power to evoke humans, along with three different forms of coercive and marginalized human labor at the silenced center of US life, is the goal of chapter 1, "The Global Biography of Early American Coral." This chapter focuses on three of the most common coral objects encountered by nineteenth-century Americans. It presents images and descriptions of these objects, histories of their uses and meanings in multiple cultures, and accounts of the labor required to procure and produce them. The chapter thus introduces readers to the most important raw materials, so to speak, that so many nineteenth-century US writers variously drew on and transformed into the political or politically relevant insights that constitute the core of this study.

Chapter 2, "'Labors of the Coral,'" explores nineteenth-century US fascination with "coral insects" and explains how and why people came to imagine these animals as "workers," and specifically as workers who labored under conditions that strikingly evoke human chattel slavery. Returning to Sigourney's "The Coral Insect," I place the poem within a much longer tradition of seemingly ephemeral reflections on reef-making between the 1820s and 1860s. During this period, visions of apparently collective coral "labor" repeatedly normalized coercive labor relations as the necessary foundation of the strongest structures—while simultaneously eliding the fact that Black enslaved persons were the human workers performing that type of labor in the US. That elision is precisely what later writers, speakers, and activists would come to challenge through coral. The chapter thus establishes the centrality of coral to the imagination of human labor, but it also lays the groundwork for the alternative and more hopeful modes of political thinking with coral that also emerged during the period.

Chapter 3, "Fathomless Forms of Life," turns to a different understanding of coral insects and coral reefs, one that originated in science and suggested that coral refuses easy classification or prediction. Americans were fascinated by older taxonomic debates about whether coral was animal, vegetable, or mineral, and by more recent studies of coral reefs by Darwin and others who sought to explain how, why, and where reefs grow from a profusion of interacting forces. US writers variously transformed that fascination into new and politically relevant claims, all of which emerged from coral's popular status as a fathomless form of life. Imagined in this way, coral especially appealed to writers seeking to challenge and propose alternatives to the period's obsession with measuring and ranking different groups of humans and then predicting their political value and social role on the basis of visible form and features.

Chapter 4, "Coral Collectives," examines nineteenth-century US writers and activists who drew on the unique properties of coral polyps and reefs to imagine, and sometimes enact, more just alternatives to a society that extracted the labor of marginalized groups while appearing to sustain the collective. These writers and activists include white US women as various as Stowe, Sigourney, feminist philosopher and university lecturer Ellen M. Mitchell, Boston adoption reformers Anstrice and Eunice C. Fellows, and historian and translator Elizabeth Wormeley Latimer. And they include Black men and women, from James McCune Smith to Frances Ellen Watkins

Harper, who drew on coral to point out the limits of white collectives and call for transformation of the political arrangements that subjugated Black persons long after the end of slavery. The power of coral both to illuminate coercion and model a more genuinely just collective is nowhere more apparent than in actual nineteenth-century charitable societies that named themselves after coral, such as Cleveland's Coral Builders Society, a Black literary society that supported Saint John's African Methodist Episcopal (AME) Church during the 1890s.

Chapter 5, "Red Coral, Black Atlantic," traces a long history of red coral ornaments across the Atlantic world—from Senegal, Guinea, and Nigeria through Jamaica and New Orleans—and then uses that history to reinterpret US novels and stories of slavery and race as unlikely sites of knowledge about the Black Atlantic diaspora. Black women within African and African diaspora communities across the Atlantic world wore red coral as a "Black Atlantic ritual." That ritual shaped US fictions of race and slavery as various as Stowe's *Uncle Tom's Cabin* (1852) and Chesnutt's "Her Virginia Mammy" (1899). By reading these works in light of that ritual, we can perceive that they point us toward sources of Black identity that exceed the frameworks of slavery and other forms of colonialism. Read through red coral, these works might also encourage a form of epistemic humility that must be the foundation of more just racial relations.

Collectively these chapters affirm that nineteenth-century Americans were quite capable of complexity when thinking about labor and the ideologies that rationalized labor. They delighted in acknowledging the perils, contingencies, and irresolvable quandaries attending any effort to categorize, fix, and determine. They imagined better political models than a country built on coercive labor. And yet their intellectual capaciousness and awareness too rarely translated into more just practices toward marginalized subjects. What, then, is the value of recovering that past awareness now? This is the question taken up more fully in the coda, "Coral Temporalities," which reflects on the value of a historically oriented environmental humanities.

Coral reefs are disappearing. Imperiled by human industry—and by some of the same forms of industrial violence that produced slavery and other colonial exploitations—reef ecosystems are collapsing. We already know that coral's death will endanger human lives and will have other profound ecological consequences. But the loss of coral also means the loss of vital ways to *imagine*

human life and society. As we lose coral, then, we are losing a material that is not only biologically crucial but also conceptually indispensable to us. To grasp the extent and stakes of that loss, we must return to a time when coral was everywhere; when everyday encounters with coral were encounters with worldwide, oceanic circulations of knowledge, bodies, and labor; and when, consequently, seeing coral could be a matter of seeing oneself and one's polity anew.

# The Global Biography of Early American Coral

Before 1900 most of the world's coral came from the Mediterranean Sea, where it was extracted by skilled coral fishers, taken to coral workshops in Italy for processing, and sent from there to global ports, markets, and trade routes. In the early US some coral also came directly from the Caribbean, brought home by tourists and naturalists, and some came from the Pacific, brought by scientists studying the biology and geology of coral on government-sponsored research trips such as the US Exploring Expedition (1838–42). Ideas about coral came to the early US from all these places, along with many other locations in which coral grows or holds significant cultural value, such as parts of coastal Africa belonging to present-day Nigeria, where coral retains a complex set of meanings grounded in Indigenous value systems and European trade practices dating to the late fifteenth century.

The cultural biography of coral in the early US is thus a global biography.[1] It comprises information about coral forged across centuries and cultures around the world—information about coral's mythical and biological origins and nature, its growth into massive islands and continents, its various cultural uses and meanings, and the routes and practices of production, labor, and trade that brought raw coral and more finished coral objects into early US life. But how exactly did coral get from reefs growing on faraway seafloors to the early US in the first place? Where were most Americans likely to encounter coral or representations of it? What were the most common coral materials circulating through early US life and letters? And how did different groups of Americans become familiar with different strands of coral's biography that rendered coral a rich material and metaphor for thinking through contemporary political questions relating to labor—and to slavery, wage labor, and women's reproductive and domestic work in particular?

The answers to these questions constitute the necessary groundwork for the history of US political thinking with coral that this book illuminates. Yet these questions are difficult to answer for several reasons. First, there are very few cultural or economic histories of coral that approach the topic on a global scale, considering its many uses and meanings in multiple cultures over centuries. Second, though there are significant histories of coral's uses and meanings in particular Mediterranean cultures, many (though not all) of these are in French or Italian and so remain unknown or less accessible to scholars working primarily with sources in English. Third, there is no cultural history of coral that centers on the uses and meanings of coral in the early US, and relatedly, most scholarship that touches on the topic tends to emphasize coral's significance as a talisman (a meaning that dates to the classical period within certain European cultures), thereby occluding the many additional cultures and practices that also shaped the biography of coral in the early US. Accordingly, I have assembled the concise global biography of early American coral below by drawing on information tucked away in an array of disparate early historical, literary, reference, and visual texts that circulated through the nineteenth-century US and then supplementing that information with facts drawn from a number of scholarly sources on the history of coral in English, French, and Italian.[2] I have also relied on the methods of scholars working in material cultural studies, and particularly at the intersection of race and material culture among African and African diaspora communities in the Atlantic world.

The chapter proceeds by case studies of the three most familiar coral objects to early Americans: red coral jewelry, the coral and bells, and coral reef specimens. For each object, I present color images; explain where and how Americans encountered and used these objects; and explore various early US texts that illuminate the global circuits of human labor—from harvest to production and trade—that brought coral from the seafloor to the US while endowing it with the power to evoke the polity's vexed reliance on different groups of human laborers closer to home.

## RED CORAL JEWELRY

Early American portrait painter Joshua Johnson depicted Emma Van Name of Baltimore thoroughly ornamented in coral (1805; fig. 1.1). The child wears a necklace of red coral beads, matching coral bar pins at each shoulder, and a silver "coral and bells"—toy, teething device, and talisman—on a long gold chain around her neck. Coral appears in a relatively high proportion of portraits by Johnson, an icon of American folk portraiture and the earliest known

FIG. 1.1. Bedecked in coral: This Baltimore child wears a necklace of red coral beads, matching coral bar pins at each shoulder, and a silver "coral and bells" on a long gold chain around her neck. Joshua Johnson, *Emma Van Name*, 1805 (The Metropolitan Museum of Art).

professional Black portrait painter working in the early US. Born in the West Indies and active in Baltimore between the 1790s and 1820s, Johnson painted local clients and their families, who were mostly abolitionists.[3]

Yet interest in coral ornaments was not confined to any particular political ideology, geography, race, or class in the early republic. Portraits by American

FIG. 1.2. A red coral necklace: In this portrait of three sisters, daughters of African American real estate investor Samuel Copeland of Massachusetts, the child at left wears coral. William Matthew Prior, *Three Sisters of the Copeland Family*, 1854 (Museum of Fine Arts, Boston).

painters as various as Thomas Sully, Clarissa Peters Russell, Charles Willson Peale, George Caleb Bingham, and others feature white women and children from all regions of the US bedecked in coral. And in William Matthew Prior's portrait of the children of African American real estate investor Samuel Copeland of Massachusetts, one of Copeland's daughters wears a coral necklace (1854; fig. 1.2).

Those persons who appeared less frequently in the period's portraiture also wore coral. Travel narratives, novels, letters, and other texts establish that coral jewelry belonged to free Black persons of various classes and to enslaved Black persons who sometimes received coral as a gift from the master's family. The relative affordability of such jewelry also made it accessible to less privileged white families. Coral ornaments came in many forms in addition to necklaces and bracelets, including pins, earrings, rings, and tiaras, and in multiple styles, from the smooth, round beads of Johnson's portrait to the fringed and angular pieces in Prior's portrait. Almost all these ornaments were made of precious red coral, *Corallium rubrum* in Latin. Very few were merely ornamental in meaning.[4]

One important reason so many women, white and Black, fastened coral to children's bodies is that coral was long believed to hold talismanic and apotropaic powers, warding off a variety of maladies both physical and spiritual, as I will explain more fully in the next section of this chapter. These powers date to antiquity, and they persisted well into the nineteenth century, a period when infant mortality remained a tragically common occurrence, which is why coral beads remained a popular christening gift. Yet by the time coral was worn by the children who appear in early American portraits by Johnson, Prior, and others, red coral ornaments had also acquired an array of additional cultural meanings among different groups of people.

To many living in or passing through the nineteenth-century US, even the smallest piece of red coral jewelry brought to hand the labor—human and nonhuman—that produced coral. An essay in *Godey's Lady's Book* titled "Coral Reefs" (1858), for example, opens with these words: "Trusting that many of our lady readers who wear coral ornaments would like to know something of their formation, we publish the following."[5] The essay offers a detailed description of reef formation—drawn from European travel narratives and Darwin's account of coral—that fashions "coral ornaments" as labor itself congealed. Each piece is literally the result of the "wonderful labors" of polyps who "die . . . to increase" the reef and also the product of an arduous process of human labor; the dangerous work of Mediterranean coral fishing is described in the final paragraph.[6] And while "Coral Reefs" reached an audience of mostly "lady readers," other genres of writing—including popular natural histories for children and Evangelical pamphlets for all—taught additional segments of the population to connect the familiar red coral jewelry circulating through daily life to the life-consuming labor of polyps and persons.[7]

Yet the power of coral ornaments to evoke particular groups of laboring bodies, nonhuman and human, is perhaps rendered nowhere as vividly and compellingly in a single text as it is in "The Story of a Coral Bracelet" (1861) by West Indies-born British writer Sophy Moody.[8] Published transatlantically in Moody's collection of children's tales, this autobiography of a red coral bead skillfully weaves together details from multiple sources of knowledge about coral's origins and growth; its natural, cultural, and economic histories; and its journey from seafloor to parlor. Moody's story thus allows us to deepen our knowledge of several intersecting histories of different forms of labor that alternately produced and imaginatively inscribed early American coral, such as that adorning the children of Johnson's and Prior's portraits (fig. 1.1 and fig. 1.2). These include polyp labor, Mediterranean coral fishing, coral manufacturing by Italian "coral workers," the global coral trade, white women's reproductive work, and transatlantic slavery.

"I am not originally a denizen of earth," for I was born "far down below the bright blue waters of the Mediterranean Sea," begins Moody's coral bead narrator, addressing her London audience of other "inanimate things . . . endowed with speech," Christmas presents that decide to narrate their biographies to pass the time on Christmas Eve.[9] The bead proceeds with a description of her birthplace, "the crystal depths" of "the Straits of Messina"—the narrow stretch of water between the eastern tip of Sicily and the western tip of mainland Italy—where she passed a blissful childhood, marred only by "sad stories" of "outrages" against fellow corals who were taken by "the hand of the spoiler."[10] Nonetheless, with "the happy forgetfulness of youth," she hoped for "a bright future," and the polyps constructing her body "watched with loving pride" as her "smooth limbs, so brilliantly red in colour," expanded the reef to which she belonged.[11] The coral was "happy, unconscious of what [its] tenth year should bring forth," yet "the spoilers' trade is, alas! an organized one," and "accordingly, no sooner have we attained the perfection of our beauty, than our wretched parents know they will lose us forever."[12] The coral then shudders at recalling the arrival of men in boats who tore her from her reef home, placed her aboard a ship laden with other corals, and took her to a faraway port.[13] There, corals from all over the world are "carried on shore," sorted by size, and subjected to "new pains and sufferings," their bodies "tortured" into jewelry and other objects that "suit the caprice of man," for in this world our "value principally depends on our colour."[14]

Upon her painful transformation into a bracelet, the coral passed "quickly from one trader's hands to another" before reaching a London shop where she was recently sold as a Christmas present.[15] "I know not what my future destiny will be," she laments, before concluding with a reflection on her name, earlier given as *Corallium rubrum*, the Linnaean classificatory name for red coral. "Ere I say farewell, I should tell you of another name which I bear, given to me, I am told, from a Greek poet's legend of the Coral's birth."[16] The coral then relays the Ovidian myth—as translated by Renaissance writers—that coral emerged when the blood of the Gorgon Medusa flowed into a "plant growing upon the sea-shore" (seaweed), transforming it into a "strange, inexplicable plant," brittle and red, that sea nymphs found a "mysterious thing," called ever afterward "*Gorgona Nobilis*."[17]

Among the many global histories of human labor informing the coral's personal history, Mediterranean coral fishing is the most explicit, experienced by the narrator as a "violent" process of being suddenly "entangled," "ensnared," and then "rudely broken off from our parent stem and carried up in our hempen coils to the surface of the water."[18] These details, along with the coral's birth in the Strait of Messina, suggest that Moody's main source of information on coral fishing is a travel narrative by Lazzaro Spallanzani (1729–1799) that provides incredible detail on the economic history of coral through the late eighteenth century, from the circumstances of its harvest, to its production, to the global coral trade. Translated into English and published in London in 1798 as *Travels in the Two Sicilies*, this text was widely read in Europe and the US.[19] Its broad circulation and clear influence on Moody's tale make it worth our while to examine for details about some forms of labor that many early Americans imagined when they encountered coral.

Spallanzani, who draws his account of Mediterranean coral fishing from "the relation of the fishermen themselves," accurately describes the fascinating, and sometimes gruesome, details of a practice pursued throughout the Mediterranean Sea, though particularly in the western basin along the coasts of North Africa and France, from the mid-seventeenth through nineteenth centuries.[20] Performed almost exclusively by men from poorer French and Italian coastal villages bordering the Mediterranean, the labor of coral fishing was, by all accounts, a highly skilled form of work relying on knowledge passed down within families from one generation to the next.[21] It was also an "extremely laborious and fatiguing" enterprise that exposed the laborers to numerous and sometimes life-threatening dangers.[22] Each year in the spring and summer, hundreds of coral fishers, known in Italy as *corallini*, went to sea in specially crafted, small rowboats called *corallina* that held a crew of about seven. For retrieving coral, the boat was equipped with the *ingegno*.[23] Spallanzani's description of this device—which Moody's coral calls "the clumsy, cruel apparatus"—comes from firsthand observation and the historic illustrations in *Histoire Physique de la Mer* (1725; fig. 1.3 and fig 1.4), a natural history of the sea by Luigi Ferdinando Marsigli.[24] The *ingegno* consisted of two crossed wooden planks, each about thirteen to sixteen feet long. To each of the four ends of the cross, and sometimes at the center, were fastened groups of nets of about thirty feet in length. A rope was attached to the place where the planks crossed, and a heavy stone affixed below as a weight.

Once the *corallina* reached the general location of the intended reef, the coral fishers exerted incredible strength and skill to use the *ingegno*. Holding tightly to the rope, they would throw the device from the stern of the boat, lower it as much as one thousand feet to the seabed by way of a geared hand-winch, and then begin the tedious and tiring labor of moving the apparatus around to make contact with the reef, sometimes diving to make adjustments. At this point the labor of ensnaring began as one of the fishermen worked the rock, pulling the *ingegno* back and forth, often in turbulent waters, and sensing by experience

FIG. 1.3. Coral fishing: Mediterranean coral fishers went to sea annually in spring and summer in small rowboats called *corallina* that were equipped with a mechanical device called an *ingegno* for wrenching coral from the sea floor. Coral fishing was a highly skilled and dangerous job. Louis Ferdinand comte de Marsilli, *Histoire Physique de la Mer*, 1725. Figure 109.

FIG. 1.4. The *ingegno*, which Sophy Moody's coral bracelet narrator calls "the clumsy, cruel apparatus," was cast onto the reef and then maneuvered back and forth to ensnare the coral. Louis Ferdinand comte de Marsilli, *Histoire Physique de la Mer*, 1725. Figure 110.

how to entangle the living coral as fully as possible in the nets. The crew then winched the heavy cross upward by its main cord, breaking the coral from the reef, and hauling it aboard the boat where they disentangled the coral from the netting and prepared to lower the *ingegno* again. On any given trip this process was repeated five to six times across a workday that typically spanned fifteen to eighteen hours and included risk of all manner of perils and accidents, some of them fatal. Spallanzani describes "violent" winds and currents that, when combined with the weight of the *ingegno*, could overturn the boat.[25] The cord connected to the *ingegno* could snap, pulling the fishermen into the sea. And on the way to and from reefs, coral fishers could be attacked, captured, and even potentially enslaved by corsairs.[26]

Nonetheless, "through the long, bright summer days the cruel work goes on, until September, when [the coral fishers] return to their homes laden with the rich spoil which they have snatched from ours," reflects Moody's coral bead.[27] Most coral fishermen brought their bounty to the Italian coastal town of Livorno, which was the world's center of coral manufacturing and trading beginning in the mid-seventeenth century. Each fall, in a seasonal market called the "coral fair," these fishers unloaded their coral along Livorno's docks, where it passed through the hands of merchants and then to many nearby "coral workshops" where it was, in the words of the coral narrator, subjected to "new pains and sufferings."[28]

In these workshops—many of them still in operation today in Livorno and other places—"coral workers," as they are called in many contemporary sources, commenced "grinding, drilling, and polishing the coral," a few of the steps in a time-intensive labor process carried out by a workforce largely comprising women.[29] Thus was raw coral transformed into cameos or beads of various shapes and sizes; the beads were sometimes then, as the coral narrator relates, "strung into necklaces and bracelets like myself."[30] Finally, coral workers prepared these objects for packing aboard ships bound for ports all over the world, including London, where the coral herself ends up for sale in "a shop in Charing Cross."[31]

Moody's "Coral Bracelet" thus illuminates one global route that much coral jewelry took from the Mediterranean seafloor to shop windows in Europe and the US via the labor of coral fishers, coral workers, merchants, and other participants in the coral trade. Yet by telling the history of that trade from the perspective of the coral, Moody simultaneously illuminates a very different, yet connected, global route of labor that coral powerfully evoked: the transatlantic slave trade, in which Mediterranean red coral also played an important role.

It cannot escape the reader that the coral narrates her story in the form of a slave narrative, beginning with the genre's most conventional opening, a description of the narrator's birthplace.[32] The coral's personal history then closely follows that of a Black woman kidnapped from Africa and forced into slavery in the Americas: she is stolen from her home by "cruel" traders who arrive in boats, valued for beauty and color alone, subjected to physical torture, packed aboard a ship, taken on an oceanic voyage, and transformed into a commodity for sale. Furthermore, these experiences have given the bead a keen understanding of slavery's particular peril for Black girls, which she describes in words that startlingly echo those of Harriet Jacobs in *Incidents in the Life of a Slave Girl* (1861), published during the same year as Moody's tale. Just as Jacobs explains of the "female slave" that "before she is twelve years old" her "beauty . . . will prove her greatest curse," the bead laments that "no sooner have we attained the perfection of our beauty"—around age ten—"than our wretched parents know they will lose us forever" to "the spoilers' trade."[33]

As a formerly enslaved Black girl, Moody's narrator may partly embody what scholar of African American literature and culture Nazera Sadiq Wright has identified as a politically powerful nineteenth-century "trope of black girlhood" that furthers an antislavery agenda.[34] As both coral and formerly enslaved Black girl, she also embodies a very specific relation between red coral and the body of an enslaved or formerly enslaved Black person. This relation took a number of different forms in European and American print and visual culture before 1900. An investigation of this relation, its origins, and its history, links Moody's tale more closely to transatlantic slavery and also to certain practices and meanings of coral that developed within specific cultures of Africa in tandem with the slave trade and then persisted long beyond it, both there and in parts of the African diaspora.

The relation in question appears frequently as the figure of a black girl or woman holding or (more commonly) wearing red coral. Several early modern European portraits of free or enslaved Black persons throughout the Americas feature Black women in red coral necklaces, and one particularly well-known instance is Dutch artist Albert Eckhout's painting of free persons of African descent in colonial Dutch Brazil, *African Woman and Child* (1641; fig. 1.5).[35] The figure of a black female child grasping raw coral (1760; fig. 1.6), or arrayed in a red coral necklace (1750; fig. 1.7), was a familiar feature of eighteenth-century ornamental porcelain sets of "the four continents"—allegorical representations of Europe, Asia, Africa, and America—that graced mantelpieces and dinner tables in the homes of wealthier Europeans and Americans.[36] And in a number of nineteenth-century US novels of slavery and race—analyzed in this book's final chapter—Black or mixed-race Black girls or women covet, wear, or deploy red coral jewelry of various kinds, such as the earrings, necklace, and bracelet adorning Topsy in US painter J. A. Bingham's illustration of a scene in *Uncle Tom's Cabin* (1857; fig. 5.5).[37]

What are the potential origins of this widespread relation? According to many historians of European and American art, it stems from the Mediterranean coral fishery that fueled the global coral trade. Since coral grows in

FIG. 1.5. Allegorical coral: While free persons of African descent in colonial Dutch Brazil might have actually worn red coral necklaces, this portrait may be more allegorical than documentary, considering the European artistic tradition of portraying the continent of "Africa" as a woman wearing or holding red coral. Albert Eckhout, *African Woman and Child*, 1641 (National Museum of Denmark).

FIG. 1.6. Holding coral: The figure of a black female child grasping raw coral (or wearing it, as in Fig. 1.7.) was a familiar feature of eighteenth-century ornamental porcelain sets of "the four continents," allegorical representations of Europe, Asia, Africa, and America, that graced mantelpieces and dinner tables in the homes of wealthier Europeans and Americans. Frederick [Friedrich] Elias Meyer, "Putti as Africa and Europe," Meissen Manufactory, 1760 (Detroit Institute of Arts, USA © Detroit Institute of Arts / Gift of Mr. and Mrs. Jacques Linsky / Bridgeman Images).

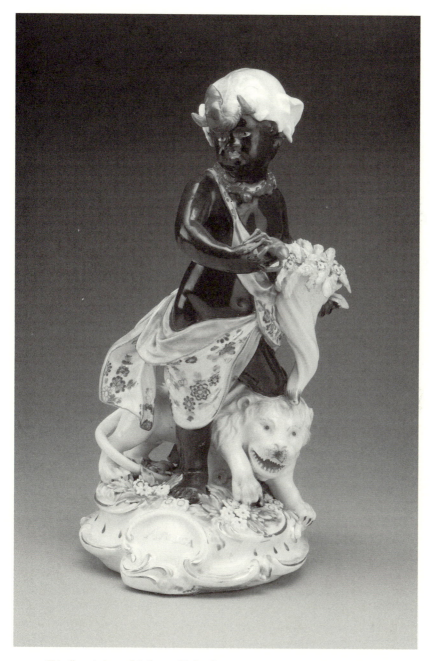

FIG. 1.7. This allegorical porcelain figure of "Africa" wears a red coral necklace, in keeping with a long-standing European artistic tradition. Allegorical Figure of Africa, Derby Porcelain, ca.1750–1848 (Art Institute of Chicago).

abundance along Africa's Mediterranean shores, the relation—particularly
when it appears as the figure of a black woman wearing or holding coral—
signifies the African continent's "contribution" of coral to the global marketplace, given the abundance of Mediterranean red coral growing along the
North African coast. "Africa" thus wears coral for the same reason that "Asia"
frequently holds a vase filled with spices: both belong to a European representational tradition of the continents that dates to antiquity and originally
figured their commercial identities and exchanges. A famous codification of
that tradition appears in Cesare Ripa's *Iconologia* (1603), an illustrated volume
of moral emblems that remained the standard handbook among European
artists for about two hundred years. This text, published in English in 1709,
illustrates "Africa" as a woman wearing a coral necklace and coral branch pendant, while the accompanying passage encourages artists to allegorize Africa
as a woman wearing "a Necklace of Coral; and Pendents of the same, at her
Ears" (fig. 1.8).[38]

Yet this origin story obscures two other deeply entwined cultural practices
that may also have contributed to the genesis of this convention and that certainly informed its reception and interpretation until well into the nineteenth
century. These practices are the European trade of red coral for African persons
to enslave in the Americas, and the Indigenous African value systems that made
red coral valuable enough to exchange for persons in the first place. Both practices originated during the early modern period in particular locations along
the western coast of Africa, from present-day Senegal to Angola.

We can begin to uncover these two relatively obscured histories of the convention by joining a standard European and American art historical account of
it to the work of art historians, anthropologists, cultural historians, economic
historians, and other scholars of Africa and the Black Atlantic diaspora who
have explored the role of red coral in the transatlantic slave trade. This work
strongly suggests that we read early representations of Black women from
Africa wearing coral more literally: it indicates that many artists and writers
may have portrayed "Africa" as a woman in red coral jewelry because women
(and men) who lived in specific parts of coastal Africa in fact *wore* such jewelry
themselves.[39]

That practice is widely documented, sometimes visually, in numerous published European travel narratives of Africa.[40] A particularly notable example is
Louis de Grandpré's *Voyage à la côte occidental d'Afrique* (1801), which features
engravings of an Angolan man and woman wearing red coral jewelry, and relates that their culture values "corail rouge" among the most prized personal
adornments, on par with the status accorded to diamonds in many European
cultures (fig. 1.9).[41]

*AFRICA.*

di effa anco alla zona torrida ; onde gli Africani vengono ad effere natu-
ralmente bruni , & mori .

Si fa nuda , perche non abonda molto di ricchezze quefto paefe.

La tefta dell'elefante fi pone , perche così fta fatta nella Medaglia de
l'Imperadore Adriano , effendo quefti animali proprij de l' Africa , quali
menati da quei popoli in guerra , diedero non folo merauiglia : ma da
principio fpauento à Romani loro nemici .

Li capelli neri , crefpi , coralli al collo , & orecchie , fon ornamenti loro proprij morefchi .

Il ferociffimo leone , il fcorpione , & gli altri venenofi ferpenti , dimo-
ftrano che ne l'Africa di tali animali ve n'è molta copia , & fono infinita-
mente venenofi , onde fopra di ciò così diffe Claudiano .

*Namq;*

FIG. 1.8. "Africa" in coral: Cesare Ripa's renowned handbook of emblems instructs artists to depict "Africa" as a woman wearing "a Necklace of Coral; and Pendents of the same, at her Ears," partly signifying
the abundance of red coral that grew along the continent's Mediterranean coast. Red coral also played
an important role in the political, economic, and cultural practices of numerous Indigenous African
cultures. Cesare Ripa, *Iconologia*, 1603.

Dessiné d'après nature, par G. P.                                                    Godfroi. Sc.

Princesse née de Malembe.

FIG. 1.9. Louis de Grandpré's travel narrative of Africa features engravings of an Angolan man and woman wearing red coral jewelry and relates that their culture values "corail rouge" among the most prized personal adornments. *Princesse née de Malembe [graphic] / Dessiné d'après nature, par G.P.; Godfroi sculpt.* In Grandpré, L. de (Louis), 1761–1846. *Voyage à la côte occidental d' Afrique* (Paris: Dentu, imprimeur-libraire, Palais du Tribunat, galeries des bois, no. 240, an IX, 1801).

Red coral beads in particular were (and in some cases still are) a common feature of dress, not only in Angola but also in other cultures of Africa, where they have historically served many purposes in addition to ornamentation. As Ogundiran observes, in many such cultures beads of various types were "commodities with complex cultural biographies" and social values that were "politically and culturally mediated."[42] That coral beads sometimes served as such is evident from their role in the Kingdom of Benin, which flourished from the eleventh to the late nineteenth centuries in a coastal region of present-day southern Nigeria. There coral beads in particular became what Ogundiran might call highly valuable "objects of social distinction, power, and political capital," in large part because of an oral tradition that dates to the late-fifteenth-century arrival of Europeans in the region and links the beads both to the Kingdom's political legitimacy and to transatlantic slavery.[43]

According to this tradition, red coral beads were introduced to the kingdom by Oba Ewuare, ruler from 1440 to 1473, who traveled from Benin City to the seacoast and then to the undersea palace of Olokun, deity of waters, who dressed entirely in a woven regalia and crown of red coral beads.[44] As the son of a divine king, Ewuare was able to wrest a portion of the sacred beadwork from Olokun and wear it as his own royal attire, thereby establishing the Oba's palace as the earthly counterpart of Olokun's. This origin traditionally endowed the beads with tremendous political and spiritual power: those belonging to the Oba had "the power of àṣẹ, that is, whatever is said with them will come to pass," and all coral beads in the kingdom were said to belong to the king.[45] During an annual coral feast that many Edo people continue to celebrate, "all the beads of the kings, chiefs and royal wives are gathered together" and sanctified with blood that renews their otherworldly powers.[46]

The manifold powers that this culture and others attributed to red coral rendered it valuable currency for many kinds of goods. In coastal areas of present-day Senegal, Guinea, Liberia, Ghana, Benin, Nigeria, and Angola, those "goods" included persons.[47] This use of red coral had been introduced by Portuguese traders during the late fifteenth century. It was then perpetuated by British, French, and Dutch traders for over three hundred years.

A number of published and widely circulating early modern European texts produced in support of the slave trade explicitly record and recommend the exchange of red coral for Black persons. Dutch writer Olfert Dapper's massive and highly influential tome on Africa, published in Amsterdam in 1668, includes "root Korael" in a list of materials that local slave traders at Ghana eagerly accept from Dutch traders in exchange for persons.[48] "[C]oral, large, smooth, and of a deep red," is among the most "preferable goods" with which to purchase persons in "Whydaw" (Whydah, in present-day Benin), recorded Thomas Phillips, captain of a British slave ship that brought human cargo from Africa to Barbados during the eighteenth century.[49] A widely used dictionary of trade and commerce, Chambon's *Guide du Commerce* (1777), records red coral among the sixteen most valuable items that should constitute the cargo of any ship destined for the "traite des noirs," particularly along the coast of Guinea where the locals regard coral as among the most precious natural materials, and coral necklaces as a particular status symbol.[50] In 1823 French Captain J. F. Landolphe records red coral among the many goods for which the prices of persons are fixed along the coast of Nigeria.[51] And in the sensationalist and unapologetic *Adventures of an African Slaver* (1854), slave trader Théodore Conneau recounts that during the 1830s and 1840s even a small quantity of "Venetian beads of mock coral" would send hundreds of Liberians into New World slavery.[52]

It bears repeating that the exchanges of coral for persons recorded in these and other texts would not have been possible without the complex Indigenous African value systems that endowed red coral with tremendous worth. This fact warns us that the figure of a black woman in red coral did not exclusively endorse or evoke the slave trade; among some who encountered this convention, or even wore coral themselves, other meanings also persisted.

Enslaved and free Black women in certain parts of the African diaspora, for example, chose red coral jewelry as a preferred personal adornment, as a number of letters, travel narratives, natural histories, and images record.[53] And as I will discuss at greater length in chapter 5, these women were not citing the transatlantic slave trade alone but also the Indigenous African value systems that persisted (and continue to persist in many cases) as important sources of their identity.[54]

Defiance of slavery's reduction of persons to goods or specimens for trade, then, is also frequently encoded in red coral, and broadcast by the choice of Black women to wear it. The histories of both that violent reduction and the tacit refusal of it were alternately present in early American coral, there to be mobilized to radically different ends. If coral could speak to early Americans of its life, then it makes sense that it would do so in the form and conventions of a slave narrative. Each small piece might tell of worldwide circuits of labor and pain, along with the base economic motives driving the slave trade, while also—or alternatively—affirming the persistence of identities forged outside of colonial frameworks and sight lines. And coral could also speak of other groups of people who, while coerced by an economy and culture founded in slavery, experienced markedly different forms of exploitation.[55]

## Coral and Bells

On her journey through the latter stages of the global coral trade, Moody's coral bead learns that jewelry is not the only object that humans fashion from raw coral. For those pieces of coral who are particularly large and smooth, "a most singular destiny is appointed. Set in a strangely-shaped ornament of silver or gold, which is encircled by tiny bells," they become "an infant's toy . . . to soothe pain when rubbed upon its aching toothless gums!"[56] Moody's "strangely-shaped ornament" is the coral object at the center of Johnson's portrait of Emma Van Name (1805; fig. 1), the "coral and bells," sometimes called "whistle and bells," that US women—both Black and white—would hang around an infant's neck or waist on a ribbon or chain.

During the eighteenth and nineteenth centuries the coral and bells was a popular gift for babies of all genders in more affluent, usually white, families. Often presented at christenings, it frequently became a family heirloom passed down from one generation to another. Such is the case with one of two such objects owned by Henry Wadsworth Longfellow (1852; fig. 1.10). In 1852 Longfellow gave it to his niece, Mary King Longfellow, who in turn gave it to a cousin's granddaughter in 1935. The object is an exact copy of one that Longfellow had earlier given to his first child, Charley, in 1844. It may even appear in the poem "To A Child" (1845), in which Longfellow's speaker tells an infant that his "coral rattle" was produced by the combined labors of polyps building coral reefs for "thousands of years in Indian seas" and South American silver miners working "far down in the deep-sunken wells / Of darksome mines."[57]

In fact, the coral portion of most early American coral and bells came from the Mediterranean Sea via the global coral trade described in the first section of this chapter. European and US silversmiths and goldsmiths imported smooth pieces of coral cut specially for use in such rattles and then fitted them to a whistle and bells made of silver—or less frequently of gold—that sometimes bore ornately engraved flowers, leaves, or scrolls. Though expensive, the item was in relatively high demand and wide use in the early US, as attested by many ads for it in major eighteenth- and nineteenth-century newspapers— including the *South Carolina Gazette, Pennsylvania Journal, Boston Gazette, New York Mercury,* and *Maryland Gazette*—and by its frequent appearance in import lists, bills of sale, and family and business account books.[58] Most early American coral and bells were imported from Europe, particularly from London, though there are numerous descriptions and surviving examples of ones produced locally by US craftsmen, such as that ordered by Martha Washington from a Philadelphia silversmith in 1791.[59]

FIG. 1.10. Teething on coral: A combined teething aid, rattle, whistle, and talisman, the "coral and bells" was a popular gift for babies of all genders in more affluent, usually white families. This one belonged to US poet Henry Wadsworth Longfellow, who gave it to his niece in 1852. "Coral and Bells," ca. 1850 (Courtesy of National Park Service, Longfellow House-Washington's Headquarters National Historic Site, LONG 27636).

According to a broad scholarly consensus, the use and meaning of this object during the eighteenth and nineteenth centuries is grounded in the same medicinal and talismanic properties that also made red coral jewelry a popular christening present.[60] In the case of the coral and bells, those powers are amplified by the fact that the object is designed for a child to grasp and even ingest. Some surviving examples bear infant teeth marks; it was believed that by chewing on the coral as they teethed, babies would absorb the curative features that Pliny the Elder documents in *Historia naturalis* (AD 77–79).[61] During a time of high infant mortality, the coral portion of the coral and bells would preserve life by protecting small bodies from physical and spiritual ills ranging from witchcraft, lightning, and epilepsy to more common sicknesses and ailments.[62]

These powers derive largely from Ovid's account of the origin of coral in *Metamorphoses* (AD 8), which endowed coral with petrifying influences. According to Ovid, Perseus frees Andromeda by slaying Medusa and then severs her head, laying it on a bed of seaweed. Thereafter,

> the fresh plants, still living inside, and absorbent, respond to the influence of the Gorgon's head, and harden at its touch, acquiring a new rigidity in branches and fronds. And the ocean nymphs try out this wonder on more plants, and are delighted that the same thing happens at its touch, and repeat it by scattering the seeds from the plants through the waves. Even now corals have the same nature, hardening at a touch of air, and what was alive, under the water, above water is turned to stone.[63]

The touch of the Gorgon Medusa thus generates coral and simultaneously imbues it with Medusa's petrifying powers, an origin story that encouraged centuries of naturalists to suggest that coral crosses taxonomic kingdoms, turning from plant to stone when lifted from the sea, and bears both transformative and healing powers.[64] These associations also rendered red coral a popular charm for guarding against the "evil eye" beginning in the medieval period—one reason that Christ wears coral in so many medieval and Renaissance portraits—and they persist to this day in the Neapolitan tradition of wearing or carrying a *cornicello* (or *cornetto* or *corno*), a small, horn-shaped object frequently made of red coral for good luck, health, and protection.

Yet in the early US the coral and bells embodied a number of other associations alongside the protective powers that Ovid ascribes to coral. For one, the object—like red coral jewelry—is deeply linked with labor, both nonhuman and human. Longfellow's "To a Child" is not the only nineteenth-century US poem that teaches children to see their coral rattle as the product of prolonged polyp work. A coral polyp itself instructs them to do so in a poem appearing in *Youth's Penny Gazette* (1848) whose speaker is a coral insect that works "without ceasing" and then explains that "the infant in arms, takes my work for his toy / And jingles his bells with great pleasure and joy."[65] The same publication printed a widely circulating poem on this item by British writer Henry Cood Watson (1842). The poem takes the form of a dialogue between mother and child: Mary implores, "O buy me a coral and bell, mamma," to which her mother responds that she will do so—but only if Mary learns about the laborers that produced the object, both the silver miner who "toils / In danger, to supply" our luxuries, and the coral insects "in the Mediterranean sea" who "labor on with patient care" for "many, many years."[66] At the same time, the poem's form and female speakers also illuminate just how deeply the coral and bells was associated with the labor of white maternity.

Deep relations between coral and white women's reproductive and child-rearing work are at the center of a number of early US portraits featuring the coral and bells. The association is especially clear in Joseph Blackburn's *Isaac Winslow and His Family* (1755; fig. 1.11). The viewer's sight line is drawn from left to right, beginning with Winslow's downward-pointing finger and continuing through his wife's right arm, which meets her left around their younger daughter's waist to grasp a coral and bells that echoes the shape of Winslow's finger and points yet farther downward to the shadowy region beneath the uplifted skirt of their elder daughter. The object's unmistakably phallic shape strongly amplifies the fertility symbolized by the daughter's skirt full of fruit.

A quite similar dynamic unfolds in an earlier painting that may have influenced Blackburn. Dutch painter Nicolaes Maes's *Interior with a Dordrecht Family* directs the viewer's eye through a line of hands that begins at left with

FIG. 1.11. Coral as fertility object: In this portrait of the Winslow family of Boston, Isaac Winslow's wife holds their younger daughter Hannah while grasping a coral and bells. Notice how the object echoes the shape of Winslow's finger and points downward to the shadowy region beneath the uplifted skirt of their elder daughter. Joseph Blackburn, *Isaac Winslow and His Family*, 1755 (Museum of Fine Arts, Boston).

the father's, passes through fruit symbolic of fertility, and culminates in a coral and bells grasped by the baby who points it upward at the mother seated at right (1656; fig. 1.12).[67]

The link between the coral and bells and women's fertility may have been even more apparent to contemporary viewers of these and other portraits containing the object because of its strong resemblance to the *mano figa* or *manu fica*, which was frequently made of coral (fig. 1.13). A hand-shaped amulet in the form of a closed fist with the thumb protruding between index and middle fingers, the *mano figa* represents sexual penetration and was used to encourage or protect fertility as early as the classical period in the Iberian Peninsula and Italy.[68]

A large number of early US portraits of white mothers and children feature the coral and bells, and it is hard to imagine that the object's links to women's fertility were unknown to early painters, subjects, and viewers, particularly considering the placement of the object within these works. In James Peale's *Portrait of Mrs. Nathaniel Waples and Her Daughter, Sarah Ann* (1817; fig 1.14), the coral shaft protrudes from the child's closed fist, evoking the sexual meaning of the *mano figa* quite explicitly.

FIG. 1.12. The painter directs the viewer's eye through a line of hands that begins at left with the father's, passes through fruit, symbolic of fertility, and culminates in a coral and bells grasped by the baby who points it upward at the mother seated at right. Nicolaes Maes (Dutch, 1634–93), *Interior with a Dordrecht Family*, 1656 (Norton Simon Museum, Pasadena, California).

FIG. 1.13. This amulet, called *manu fica*, takes the shape of a closed fist with the thumb protruding between index and middle fingers. It represents sexual penetration and has been used to encourage or protect fertility since at least the classical period in the Iberian Peninsula and Italy. Notice how its shape echoes that of the "coral and bells" (shown in Fig. 1.10). Charm, hand; manu fica. (Division of Medicine and Science, National Museum of American History, Smithsonian Institution, acquired ca. 1898–1910).

FIG. 1.14. The shaft of this coral and bells protrudes outward from the closed fist of baby Sarah Ann Waples, evoking the sexual meaning of the *manu fica* (shown in Fig. 1.13). James Peale, *Portrait of Mrs. Nathaniel Waples (1796–1819) and Her Daughter, Sarah Ann (1816–1850)*, 1817 (Philadelphia Museum of Art).

Considered as a fertility object, the coral and bells may also amplify the primary purpose of early American portraiture, which was, according to art historian Margaretta M. Lovell, to serve as "documents of the family line."[69] Portraits were the most popular form of painting in the early US, and they met the purpose of establishing the familial bloodline, verifying origins and pedigree, and keeping heritable wealth within the family. Looking again at Blackburn's portrait of the Winslow family (1755; fig. 1.11), it seems possible that the coral, in combination with Winslow's hand and self-assured expression, assures us of the linear and orderly direction of his family's bloodline—from his person and, through his wife, to their daughter, who will in turn continue it into the future.

In this way, Blackburn's painting and other family portraits including this item also materialize associations between coral and women's blood, and particularly coral and child birth, that derive from Renaissance translations and adaptations of Ovid's account of Medusa. As early as the fifteenth century, writers and artists began adding to Ovid's account the idea that coral arose specifically from Medusa's blood, as art historian Michael Cole relates.[70] In an influential fifteenth-century translation of *Metamorphoses*, Perseus lays Medusa's severed head not on seaweed but rather on a bed of "seaborne sticks of wood" that "hardened as stone does, and from the blood of the head they became vermilion."[71] While Ovid has the vegetation absorb only the Gorgon's "touch," here and in numerous other later versions the absorption of her blood turns the plants to coral, and that coral and blood relation quickly gained currency across multiple media.[72] For example, in a bronze sculpture known to nineteenth-century Americans, Benvenuto Cellini's *Perseus and Medusa* (1545)—to which Ishmael compares Ahab in Melville's *Moby-Dick* (1851)—Medusa's blood turns to coral as it spills from her head and severed neck (1545; fig. 1.15).[73] And at some point during the Renaissance, the event of coral's emergence from Medusa's blood began to be called a "birth," thereby further linking coral, blood, labor, and women. As Moody's coral bracelet bead explains, one of her many names is *Gorgona nobilis*, "given to me, I am told, from a Greek poet's legend of the Coral's *birth*" from the blood of Medusa's head.[74]

By the nineteenth century, then, the coral and bells was more than an amulet alone. In this object, coral's powers to heal and protect the body joined other historical associations that rendered the coral and bells especially evocative of white women's bodies and maternal labor. As we have seen, a significant number of early American portraits silently but powerfully mobilize all these associations through very specific placements of one particular coral object. In a final example, consider *The Copley Family*, painter John Singleton Copley's portrait of his own family (1776/1777; fig. 1.16). Of the placement of the object in the upward reaching hand of the child at left, Lovell observes that the child's gesture indicates an active "clamoring" for her grandfather's attention—a motion marking Copley's participation in a new understanding of children as individual agents.[75] But her grandfather does not notice her. Instead, he stares off into the distance, exposing his neck toward which she extends the coral that also points toward her father. Is her gesture then a kind

FIG. 1.15. Coral, blood, and women: In this sculpture, to which Ishmael compares Ahab in Melville's *Moby-Dick* (1851), Medusa's blood apparently turns to coral as it spills from her head and severed neck. Benvenuto Cellini, *Perseus and Medusa*, 1545–54 (Loggia dei Lanzi, Piazza della Signoria, Florence, Italy).

FIG. 1.16. Women and coral as keepers of the bloodline: At left, John Singleton Copley's daughter Susanna holds a coral and bells in her upward-reaching hand, pointing toward the patriarchal bloodline that she will presumably carry on one day as a mother. John Singleton Copley, *The Copley Family*, 1776/1777 (National Gallery of Art).

of lunging toward, or at least indicating or connecting to, the patriarchal bloodline that she will presumably carry on in her person? Wherever the coral and bells appears in these portraits, it evokes women's labor by pointing, whether literally or figuratively, toward the bodies that labored to perpetuate and safeguard the bloodline.

## CORAL REEF SPECIMENS

*Wonders of the Deep* (1836), a Sunday School pamphlet and natural history of the sea by US writer Margaret Coxe, tells the story of the Stanley family's summer excursion to the New England seashore.[76] One evening when rain confines the children indoors, Mrs. Stanley entertains them with a box of "various marine specimens" collected from "the shores of the Mediterranean," "the West Indian islands," and coastal New Jersey.[77] "Some of these

FIG. 1.17. George Washington kept this coral specimen in his home library. He likely acquired it on his 1751 trip to Barbados, his only journey beyond the North American mainland. George Washington's "Fan Coral" specimen (*Gorgonia ventalina*), Barbados, 1751 (Courtesy of Mount Vernon Ladies' Association).

were of a brilliant scarlet color, and branched like a delicate shrub; others were yellow and orange," and some of "snowy whiteness."[78] All are "the labors of that wonderful insect 'the coral-worm,'" explains Mrs. Stanley.[79] She proceeds with a brief account of the Mediterranean coral fishery, drawn from Spallanzani's *Travels*, and information about the growth and nature of coral as described in a variety of widely available travel narratives and natural histories.[80]

*Wonders of the Deep* points toward a number of sites where early Americans commonly encountered coral reef specimens and some of their global economic and cultural histories. Branches of coral were seen at home in curiosity cabinets, on mantelpieces, and in home libraries or studies, which is where George Washington kept two coral specimens of different species, *Acropora cervicornis* and *Gorgonia ventalina* (fig. 1.17), that he likely acquired on his only journey beyond the North American mainland, a trip to Barbados in 1751 to repair the health of his ailing half-brother.[81]

Descriptions and images of coral specimens figure in early natural histories of the Caribbean, such as Griffith Hughes's *Natural History of Barbados* (1750)—which Washington owned—and Mark Catesby's lavishly illustrated *Natural History of Carolina, Florida, and the Bahama Islands* (1729 to 1747; fig. 1.18). And drawings of coral branches appear in natural histories intended for popular or juvenile audiences, such as *Good's Book of Nature* (1834) and *Peter Parley's Wonders of the Earth, Sea, and Sky* (1837), and in guides meant to be taken along on trips to the seashore for identifying various species of corals and seaweeds while beachcombing.[82]

As nineteenth-century scientific exploration of the Pacific expanded alongside new developments in coral reef science, Americans encountered specimens of new species of coral in and beyond print. Darwin describes coral at Tahiti in *The Structure and Distribution of Coral Reefs* (1842), a work widely discussed and excerpted in the US, and soon thereafter James Dwight Dana

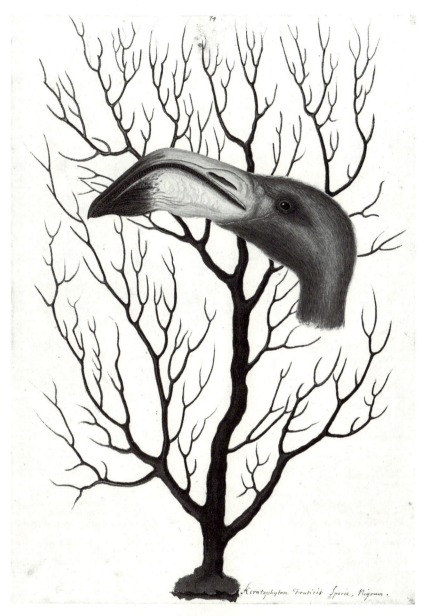

*Keratophyton Fruticis Specie, Nigrum.*

FIG. 1.18. Look behind the flamingo: This lavish illustration of a species of branching coral appeared in Mark Catesby's widely read natural history. Mark Catesby, Flamingo with *Keratophyton fruticis specie, nigrum. (Caput Phoenicopteri Naturalis Magnitudinis)*, ca. 1722–26. *The Natural History of Carolina, Florida, and the Bahama Islands*, 1729–1747, plate 74.

FIG. 1.19. Peering at coral in the nation's capital: From 1843 to 1858 specimens collected by the US Exploring Expedition were displayed publicly at a small museum in the US Patent Office Building, today part of the Smithsonian American Art Museum. Museum of the United States Patent Office, 1856 (Smithsonian Institution Archives).

published an account of the Pacific corals retrieved by the US Exploring Expedition (1838–42). These corals are illustrated in color in the *Zoophyte Atlas* (1849) that accompanied Dana's account, and one could see the specimens themselves at a special gallery in Washington, DC, that displayed the Expedition's collections from 1843 to 1858 (fig. 1.19).

The corals in this gallery, which formed the basis of the Smithsonian National Museum of Natural History, include a striking *Corallium secundum* that is illustrated in Dana's *Atlas* with vivid red branches that have faded over time (figs. 20 and 21). Emerson peered with fascination at the gallery's glass display cases of corals, as he reported in a letter to his wife, and Dana's images and descriptions captivated Melville, among other US authors.[83]

At museums Americans saw coral specimens from many parts of the world as early as the first decade of the nineteenth century. By 1802 they could visit the "Marine Room" of Charles Willson Peale's Philadelphia museum and enter a grotto of "'artificial Rock-work' to simulate a naturalistic habitat for corals and crustaceans."[84] Later in the century one did not need to own a piece of

FIG. 1.20. Reef specimen: This drawing represents a red coral reef specimen that zoologist James Dwight Dana brought back to the US, likely from Fiji, when participating in the US Exploring Expedition (1838–42). The specimen itself was publicly displayed in Washington, DC, at the Patent Office Museum. James Dwight Dana, *Zoophyte Atlas*, 1849. Plate 60, figure 1. "*Corallium secundum*, natural size."

FIG. 1.21. Dana's memorable reef specimen, bearing its original 1849 mount and label, can still be seen by visitors to the nation's capital: it lives in storage at the National Museum of Natural History. Dana's *Corallium secundum* specimen with original 1849 mount and label (National Museum of Natural History. Photo by Dr. Stephen Cairns, Research Zoologist, Emeritus, Curator of calcified Hydrozoa, Scleractinia, and Octocorallia).

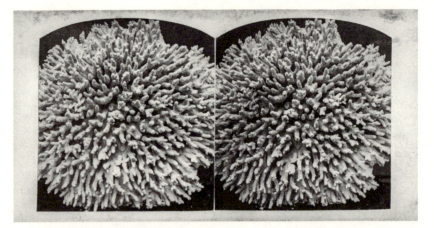

FIG. 1.22. Looking into coral at home: Coral specimens were a popular subject of nineteenth-century stereograph cards produced in sets by well-known US photographers. US author Oliver Wendell Holmes owned this one and may have peered at it through a stereoscope of his own design. "Coral Specimen" Stereocard (albumen print mounted to cardstock) owned by Oliver Wendell Holmes (Museum of Fine Arts, Boston), unidentified artist, mid-nineteenth century.

coral, visit a museum, or even leave one's own parlor to peer deeply into a coral specimen in three dimensions. Beginning in the 1850s such specimens were popular subjects of stereograph cards featuring images of coral by well-known US photographers, such as Charles Bierstadt, brother of US landscape painter Albert Bierstadt. Author Oliver Wendell Holmes owned several such stereographs with albumen prints of various coral species, which he may have viewed through the stereoscope that he designed (fig. 1.22).[85]

Wherever coral specimens appeared in the early US, they were accompanied by some of the different forms of labor that produced them. "When we admire a specimen of coral on our mantel-piece, or in the cabinet of the curious, few are aware that we see not half its beauty. We have before us a portion of a beautifully built city; but where are its gay and active inhabitants?" asks B. F. Gilman in an 1856 Sunday School pamphlet on the natural history of the sea.[86] These "inhabitants" are "the coral-working creatures," and "their workmanship . . . would cast into the shade the mightiest operations of man," Gilman explains.[87] The pamphlet captures a widespread sense that a specimen of coral brought countless nonhuman laboring bodies to hand. And that sense was arguably more visceral in the case of a coral specimen than in that of a manufactured coral object such as a bead or the coral and bells. For as pieces broken directly from the living reef, these fragments materialized what Darwin called "the soft and almost gelatinous bodies" of the "apparently insignificant creatures," and by holding them one held the "vital energies of the coral" that "conquer the mechanical power of the waves" to convert water into land.[88] Many early US descriptions of these specimens even suggest that reef fragments might somehow still be alive, or hovering somewhere between life and death.

And these fragments also encoded human labor. It was widely known that reef specimens could not be retrieved from the Pacific and the Caribbean without the skills, knowledge, and labor of local populations. Indigenous persons, enslaved Black persons, and free Black persons variously extracted coral by hand for white European and American naturalists seeking specimens, as a number of scholars have documented, for few such naturalists would have possessed the "skill, endurance, and mental resilience" that diving required.[89]

Darwin cites the knowledge of Indigenous persons on a number of occasions in his coral treatise, suggesting their importance to his development of the theory of coral reefs that would stand as one of the most significant contributions to Western coral science.[90] According to Dana, in his own later coral treatise, most of the corals retrieved by the US Exploring Expedition came from Fiji, and many by way of Indigenous Fijians in particular. During the three months that the expedition spent among the Fiji Islands "exploring the groves of the ocean," many coral "specimens were obtained by wading over the reefs at low tide" with buckets; but those growing deeper were procured "by floating slowly along in a canoe with two or three natives, and, through the clear waters, pointing out any desired coral to one of them, who would glide to the bottom, and soon return with his hands loaded, lay down his treasures, and prepare for another descent."[91] Thus, alongside Dana's descriptions and images of Pacific corals, readers caught a glimpse of some of the Fijian people whose skills, knowledge, and labor brought coral to Dana's hands.

In the Caribbean in particular diving was far more likely to be practiced by locals with experience living and swimming along seacoasts, as historian Kevin Dawson explains in a study of the "immersionary cultures" of the African diaspora.[92] Catesby himself did not dive for the coral specimens that he included in his natural history (1729; fig. 1.18), which describes a diver suffering from the sting of a jellyfish after "procuring" corals and other specimens "from the bottom of the sea six or eight fathoms deep."[93] Although Catesby does not tell us this diver's identity, he regularly depended on local labor, and Bahamians in particular "were renowned divers."[94] Local divers almost certainly also retrieved the Barbadian coral specimens that Hughes describes in his natural history in 1750, as well as those that Washington displayed in his study (fig. 1.17).[95]

Naturalists were not the only visitors to rely on local divers for coral specimens. Both during and after slavery, whites visited Caribbean resorts—often
seeking to improve their health, as Washington's brother had—and coral was a
popular souvenir. On a trip to Saint Thomas in the mid-1860s American Rachel
Wilson Moore bought a large number of coral specimens from Black divers
who "pry [corals] from the rocks," bring them to shore, and then sell "the most
delicate" and "fantastic" on "scaffolds" in front of their homes, as she records
in a published travel narrative.[96] Whites also regularly paid Black boatmen and
divers to take them to nearby reefs, sometimes called "sea-gardens," "to collect
shellfish and bits of coral," explains Dawson.[97] Once the boat arrived at the reef,
passengers were given "sea-glasses" or "water-glasses," an apparatus that one
visitor to the Bermudas in 1895 described as a "rough wooden box, without a
cover, perhaps a foot to twenty inches square, the bottom consisting of a piece
of clear glass. It is grasped firmly by the edge and held so that the bottom is
just below the surface of the water," rendering all beneath vividly and close.[98]
Passengers would then point to specific corals they desired and watch as divers pulled them from the reef, as depicted in "The Sea Gardens" (fig. 1.23), an
engraving in a popular tourist guide to the Bahamas published in 1891. At the
center of the image a white woman stands and leans forward, using her um-
brella to point toward something beneath the water, while a white man to her
left peers through a sea-glass. Surrounding the boats filled with white tourists
and manned by Black sailors, Black divers surface, and two of them lift marine
specimens above the waves to show to their eager customers. At right one diver
places a piece of coral directly into the hands of a tourist.[99]

Stark's engraving reminds us that, wherever or however coral specimens
appeared as mere decoration, they could also evoke various and differently
vexed histories of labor that so often brought coral itself to hand. While the
Black divers in Stark's print are legally free, they carry on a practice founded
in the exploitative, extractive, interlinked labors of slavery and natural history.
Consequently, they remind us that specimen collection was rarely a racially
innocent pursuit, especially in tropical places where coral grew and slavery
flourished.

By the mid-nineteenth century, however, the specific associations between
Black persons and coral specimens that emerged from slavery and colonization
also acquired new meanings that challenged the exploitative logic reducing
persons to specimens. As we have already seen, in Jamaica enslaved Black
women wore coral to preserve links to African cultures and sometimes to resist
slavery's commodifying imperative. Coral diving, too, was at times a practice
of self-determination. That fact is vividly rendered in two late nineteenth-
century watercolors by Winslow Homer that portray local Black divers bring-

THE SEA GARDENS.

FIG. 1.23. Coral for tourists: James H. Stark's popular tourist guide to the Bahamas advertised boat trips to the "sea gardens," where visitors would use "water-glasses" to look beneath the waves for interesting marine specimens that local Black divers would then retrieve. "The Sea Gardens," James H. Stark, *Stark's History and Guide to the Bahama Islands*, 1891.

ing up coral specimens from the shallow waters of the Bahamas. Upon first consideration Homer's *Sea Garden, Bahamas* (1885; fig. 1.24) seems to repeat Stark's "Sea Gardens": both are scenes of coral diving performed by local Black laborers, and there is an uncanny resonance between Homer's Black figure who stands in the water facing away from the viewer and holding a white coral branch in his right hand while leaning slightly left, and the Black figure in the foreground of Stark's engraving who does the same.[100]

Yet Homer's watercolor does more than merely document the practice of Caribbean coral diving for specimens to supply white demand. Homer visited the Bahamas in the mid-1880s on commission from *Century Magazine*, and by and large the Black subjects he portrayed there "appear within the context of their own environment—not in relation to white visitors."[101] In *Sea Garden*, the central male figure shows coral not to white customers on holiday but rather to the only other person in the painting, a Black girl who seems to admire his catch from her place on the larger boat. As he bends sideways to show off his find, the curve of his spine echoes the shape of the branching coral's spine. Perhaps this boy even holds the coral in a romantic gesture, as one might offer

FIG. 1.24. Making a living from coral: On a trip to the Bahamas in the mid-1880s, US painter Winslow Homer portrayed local Black divers retrieving coral specimens, which they could then sell to tourists. This diver seems to show off his find, possibly as a gift to the woman aboard the boat. Winslow Homer, *Sea Garden, Bahamas*, 1885 (Harvard Art Museums).

a bouquet. Alternatively, if his corals are intended for sale to white tourists, then the gesture could also signal the economic self-sufficiency of the island's free Black population. Either way, the painter centers Black Bahamians; Homer even revised the finished work by cropping several inches from the left and bottom of *Sea Garden*, with the effect of focusing our attention on the inter-action between the two figures.[102]

A similar dynamic occurs in *The Coral Divers*, with its central figure of a Black man pulling himself up from the water with a white branch of coral that engages the attention of two Black men aboard the boat (fig. 1.25). And for all of Homer's intense focus on these groups, the painter seems more interested in observing and admiring them than in making them fully intelligible to viewers, for we remain outside their world.

Perhaps these qualities of the coral paintings contributed to their exclusion from the essay in *Century Magazine* that Homer had been commissioned to illustrate. William C. Church's essay on the Bahamas, "A Midwinter Resort" (1887), trades in familiar post–Civil War tropes of the US South as a dispos-able, somewhat backward periphery of the nation—good for an occasional visit but ultimately antiquated, albeit in a somewhat charming way.[103] The racist logic of this trope is evident in Church's opening review of Bahamas

FIG. 1.25. Bringing coral from below, this diver seems ready to hand it up to two fellow workers waiting aboard the boat. Winslow Homer, *The Coral Divers*, 1885 (Private Collection).

history, which laments Spain's sixteenth-century genocide and eradication of the Lucayan, the region's Indigenous peoples, and wonders whether their "vengeful shades . . . still possess their coral islands."[104] For since that time, Church wrote, "no other race has flourished here," not even the "Conchs," Bahamian natives of European descent who sustained themselves mostly by salvaging wrecked ships, a form of labor that compromises the US economy and is thus "little better than organized robbery."[105] The figures in Homer's coral watercolors, however, are neither doomed nor enervated, and there is no haunted coral here.

Homer's images instead capture something closer to the "vital energies of the coral" that Darwin describes—coral's power to find a way forward, and to transform and develop in ways that defy human efforts to classify and predict. Nineteenth-century US authors, both white and Black, took up that meaning and transformed it to very different political ends based on the particular needs of their communities, as I document in the latter chapters of this book. In some of these writings, branches of coral evoke a history of taxonomic crisis in which coral defied the efforts of white naturalists to taxonomize and rank all of nature—even as they collected it—within one great "scale of being." In others, coral tacitly signifies the irreducibility of persons to specimens. In still others—most notably the work of Black abolitionist James McCune Smith and

Black poet Frances Ellen Watkins Harper—coral signifies the potential of Black communities to expand by sustaining one another.

Collectively these texts remind us that the meaning of coral also exceeds colonial frameworks. In the particular case of Homer's Black divers, the image captures more than a vexed history of labor. Reaching upward and toward one another, they also extend coral's defiant and transformative powers.

## 2

# "Labors of the Coral"

When US writer Margaret Fuller contemplated the beginnings of Brook Farm in Massachusetts in 1840, she imagined the founders of this transcendentalist experiment in communal living as reef-building polyps. Fuller described them to a friend as "coral insects at work," using the popular term for coral polyps.[1] Brook Farm was short-lived, foundering on a combination of financial mistakes and the difficulties of maintaining labor equity. But the analogy of coral reef to self-sustaining human collective lived on.

A coral reef, as we have already seen, gave material form to Marx's anti-capitalist vision of a less exploitative society, rising as a "mighty" structure from the communal labor of polyps who, though individually "weak," are strong in the aggregate.[2] George Orwell evokes a similar ideal in *Nineteen Eighty-Four* (1949) by way of the tiny reef fragment encased in Winston Smith's glass paperweight, which seems to belong "to an age quite different than the present one" and feels like "a compromising thing, for a Party member to have" in the novel's totalitarian world of socialism gone terribly wrong.[3] Gazing into the glass, Winston imagines that "the coral was Julia's life and his own, fixed in a sort of eternity," ever beyond their reach.[4] When the Thought Police finally come for Winston they shatter the paperweight, scattering glass and "the fragment of coral," which now appears to Winston impossibly "small."[5]

The power of coral to evoke common good through collective labor is especially apparent when writers appropriate the reef analogy to paper over the inherent violence and exploitations of colonialist projects, as some literary scholars have observed. A number of Victorian writers evoked the analogy to portray the British empire in the South Pacific as a venture of inclusive social uplift and philanthropy, Michelle Elleray shows, while others fashioned Great Britain as a "coral empire," the moral and "organic" product of many "industrious and spiritually dedicated workers," according to Ann Elias.[6] Such writers, Elleray points out, used coral to create a highly effective "cultural fable," Laura Brown's term for a story that emerges across disparate reflections on a culturally resonant material over time.[7] According to this coral fable, empires grow best by galvanizing colonizer and colonized alike in collective labor "toward a socially mandated purpose."[8]

Such examples attest that people have long imagined coral reefs as ideal models of collective self-sustenance, with polyps as tiny "workers" who labor together over time to produce and expand a reef while receiving civic identity and belonging, along with material sustenance. There is no displacement or extractive labor, for each polyp contributes and receives in relatively equal measure. Imagined this way, coral reefs have historically sponsored bright political dreams of a human society that works in the same fashion.

This chapter, however, asks what other political imaginings of human labor and laborers transpire within popular nineteenth-century reflections on reef-making, many of which seem, upon first consideration, to have nothing to do with humans at all. The labors of coral insects, imagined as the "workers" who produce and expand reefs, was a topic of widespread fascination to nineteenth-century Americans. Their imaginative reflections on these creatures can sometimes seem to participate in the same cultural fable of harmonious, collective labor that Elleray identifies. Yet in their extraordinary detail about the form of labor that coral insects perform, the relation of coral insects to the reef, and the distinct characteristics and capacities of coral insect bodies, these texts also offer another political lesson in labor: the most robust structures emerge and expand by wholly and violently extracting the labors and lives of the many for the benefit of the few.

Numerous US reflections on reefs by writers of different identities and objectives propose that a reef could not grow into an island or continent for humankind to inhabit without generations of millions of workers engaged in ceaseless, repetitive labor, from birth until death. Moreover, many Americans imagined these laborers as biologically adapted to contribute both their labor and their bodies in full to make a reef foundation suited to a less wholly corporeal form of life: all the needs of coral insects are fulfilled, or somehow cease to matter, as the reef rises from their laboring bodies, which vanish from the sight and memory of those living above the waves. I have already briefly discussed how US poet Lydia Huntley Sigourney's "The Coral Insect" participates in this conception of coral insect labor.[9] This chapter places Sigourney's poem within a vibrant tradition of nineteenth-century US writing in which an ostensibly reassuring reef analogy repeatedly gives way to an account of inexhaustible laborers whose work generates not social recognition, liberty, or belonging but rather the essential foundation on which others thrive.

We already know that nineteenth-century US visions of the common good developed in tandem with the realities of coercive labor, including the most exploitative, violent, and formative version of such labor: slavery. And as a number of scholars have shown, US print culture mediated the tensions between political ideals and realities. Yet scholarship on this topic tends to focus on representations of explicitly *human* labor and laborers and (perhaps consequently) charts a somewhat unsurprising cultural response to what historian of labor Seth Rockman calls "the persistence of coercion" within a country formally dedicated to human liberation: people either produced "social fictions" of labor that denied, erased, or minimized the extractive labor relations sustained by law, economic necessity, and cultural structures of domination, or they protested and demanded reform of those relations.[10]

By contrast, writing about coral encodes another dimension of the US political imagination of labor: a widespread, albeit tacit, acceptance of the necessity of extractive labor, alongside a casual elision of the most extractive form. Rather than trying to conceal or protest the centrality of extractive labor to an apparently thriving collective, many Americans merely normalized it through everyday reflections on coral that indulge an extractive labor logic that is strikingly continuous with one that arose to rationalize slavery, the fantasy that one form of life is wholly suited to labor for the sake of a less corporeal form. In these seemingly ephemeral coral reflections, the most robust reefs arise from "coral insects," whose labor is inexhaustible because it meets all their needs, while fusing them so fully to the reef that the body becomes inseparable from the labor product.

And white Americans in particular tended to imagine the human equivalents of these beings as the white working classes. For in almost all cases when white non-working-class writers draw an explicit analogy between coral insects and a specific group of human laborers, they single out white wage laborers as the polity's true "coral insects." In addition to a disturbing acceptance of the exploitative conditions of white wage labor, that identification speaks to a troubling disregard of the formative role of African slavery. For these writers admit the country's reliance on the reduction of human life to what Black studies scholar Ian Baucom calls "calculable matter," and then identify white wage labor as the most obvious manifestation of that reduction; they embrace the violence and extraction that arose to rationalize slavery, while failing to acknowledge the formative role (or possibly even presence) of enslaved persons.[11] Reflections on the "labors of the coral"—the title of an essay published in 1853—thus afford us new avenues to continue the project begun by scholars of racial capitalism, that of uncovering slavery's centrality to capitalism's development in the US, and understanding how that centrality was also effaced.[12]

In what follows I begin with the story of how coral insects became a metaphor for extractive labor. I then track how two early poems—Sigourney's "The

Coral Insect" and James Montgomery's *The Pelican Island* (1827)—forged a popular US conception of "coral insects" as tiny "workers" who dedicate their "labor" and bodies in full to forming the foundation for a less exclusively corporeal form of life. I next show that this conception was disseminated across US literary culture, taking on more detail about the bodies of coral insects as inexhaustible laborers suited to and inseparable from their labor product. A final section turns to writing that casually identifies white wage-laborers as the polity's true "coral insects."[13]

This chapter illuminates how coral sometimes abetted colonial thinking: it gave material form to what literary and cultural theorist Lindon Barrett calls an "elusive cultural logic that is deeply invested in making various groups visible and divisible in the first place."[14] Yet this chapter also establishes the crucial context for the following two chapters, which turn to how Americans challenged and revised that conception of coral in order to produce new accounts of reefs that would aid anticolonial thinking and practice. Put another way, by attending to the polity's fascination with the "labors of the coral," we unearth more than unexpectedly widespread consciousness of the fundamental coercions of industrial capitalism and slavery. We also begin to see that, as we lose coral now, we are losing a material that can help us perceive, question, and imagine alternatives to the polity's dependence, past and present, on persons who have never been fully included.

## CORAL INSECT LABOR

"What are Corals?" reads the title of an essay by US naturalist Elizabeth Agassiz that appeared in an illustrated periodical for children in 1869. "Before telling you what corals are," begins Agassiz, "I will tell you what they are not, because a very mistaken impression prevails about their nature. It is common to hear people speak of coral insects . . . This is a mistake. There is no such animal as a coral insect."[15]

Agassiz was by no means the only naturalist attempting to dissuade Americans from speaking of "coral insects," a project that consumed much ink and energy across and beyond the nineteenth century. The refrain "there is no coral insect," or some variation thereof, appears in a bewildering array of writings for different US audiences. The author of a popular natural history pleaded with America's youth "to remember that the animals that form coral are not insects" but rather "polyps."[16] An essay in a widely read women's periodical admonishes that it is "a very wide-spread mistake" to imagine that "little insects" make coral, for "there is no such thing as a coral *insect*."[17] Addressing readers of a weekly family magazine published in New York, naturalist Charles Frederick Holder declared in 1895 that coral insects are one of "the most remarkable and erroneous ideas regarding animal life [that] still hold in the public mind"—but this did not stop people from writing about them.[18] "There is no coral insect," a geologist felt compelled to state in 1907, while as late as 1924 the author of a treatise on corals published on both sides of the Atlantic forlornly voiced the wish that "the expression 'coral insect' may disappear from our language."[19]

The problem, according to those who objected to the phrase, is scientific inaccuracy and, specifically, the fact that when people imagined coral polyps as insects, they nearly always also imagined that polyps *labored* to build the reef. "A certain poet once announced in verse that the coral insect worked ceaselessly," Holder reflects, "and all the united efforts of naturalists for the past thirty years have failed to correct this incorrect statement."[20] The idea that polyps "worked," whether "ceaselessly" or otherwise, particularly irked America's most vociferous opponent of coral insects, zoologist James Dwight Dana, whose words provide a more detailed understanding of the issue. In an 1849 treatise on the geology of the US Exploring Expedition to the South Pacific (1838–42), Dana writes, "Very many of those who discourse quite learnedly on zoophytes and reefs, imagine that the polyps are mechanical workers, heaping up these piles of rock by their united labours." In fact, he continues, "it is not more surprising nor a matter of more difficult comprehension that the polyp should form coral, than that the quadruped should form its bones."[21] Coral, Dana proceeds to explain, is "a simple animal secretion, a formation of stony matter" that is "no more an act of labour than bone-making in ourselves."[22] Here Dana zeroes in on the issue that many naturalists had with popular descriptions of coral insects: they imply that coral making is an act of labor, and possibly even collective labor, rather than a passive limestone secretion. The proper terminology is polyp, and polyps passively secrete.

Setting aside for a moment the question of why the term "insect" readily evoked labor, it is important to understand what Dana and other naturalists were up against. The scope and scale of US interest in the lives of coral insects is astonishing, based on the various groups identified or addressed by naturalists who insisted that the coral insect did not exist. Interest in coral insects engaged a broad cross section of the American public that included people across genders, classes, ages, and racial identities; philosophers, clergymen, and public speakers; and writers of poetry, prose, and even song, for in 1856 one objector specifically accused musicians, lamenting "all that has been said and sung" of coral insects.[23]

Could it really be that this widespread phenomenon persisted across and beyond the nineteenth century entirely because of "error" or "mistake," a mere lack of knowledge about coral and reef-making, as so many naturalists claimed?[24] Admittedly it would have been difficult for many people outside of official scientific institutions to keep up with the most cutting-edge developments in coral science and terminology during the nineteenth century. At the same time, however, this was a period when science had not yet become overly specialized and when a broad and interested public eagerly read about and participated in scientific debates.[25] The growth of reefs from polyps was an especially popular topic, and clear scientific explanations appeared in numerous natural histories read by generations of schoolchildren, such as Elizabeth Agassiz's *First Lesson in Natural History* (1859), which boasts a robust reprinting record. Public lectures by coral scientists such as Elizabeth Agassiz's husband, Louis Agassiz, were well attended in the US, while popular periodicals covered theories of reef formation by Darwin and Lyell during the 1830s and 1840s.[26] These facts suggest that it would be rather simplistic to side with a chorus of naturalists who explain the persistence of coral insects as a matter of scientific ignorance alone. To understand why so many people were speaking, singing, thinking, writing, and reading about coral insects, we have to turn to these people's words.

In every publication in which Dana criticizes the popular language of coral, he accuses a single literary work of perpetuating the idea that coral is created through labor: *The Pelican Island*, an epic of nine cantos in blank verse by evangelical Scottish poet James Montgomery. Published to great acclaim in London and Philadelphia in 1827, the work describes the birth and evolution of a Pacific coral island that eventually sustains a pelican rookery. "It is not, perhaps, within the range of science to criticise the poet," Dana reflects in a footnote in his *Geology of the United States Exploring Expedition* (1849), yet "we may say in this place, in view of the frequent use of the lines even by scientific men, that more error in the same compass could scarcely be found than in the part of Montgomery's *Pelican Island*, relating to coral formations. The poetry is beautiful, the facts nearly all errors."[27] Dana later repeated his critique in the pages of *Norton's Literary Gazette* (1853).[28] And given the popularity of *Pelican Island*, Montgomery may well be the "certain poet" whom Holder also indicts for having "announced in verse that the coral insect worked ceaselessly."[29]

Of particular interest to American audiences were certain passages from canto 2 that describe the birth of the eponymous Pelican Island from the combined "labours" of "millions" of "creatures." These passages were especially widely excerpted and printed in US periodicals and anthologies, and though Montgomery himself never uses the term "coral insect" for the epic's coral-producing animals—instead calling them variously "agents," "unconscious . . . instruments," "architects," "reptiles," "creatures," "artizans," and "builder-worms"—numerous Americans identified them as coral insects.[30] In canto 2 "an island is reared by the coral insects," writes one enthusiastic reader of the epic in Philadelphia's *Album and Ladies Literary Gazette*.[31] An equally effusive review printed in Baltimore's *North American* and New York's *Albion* excerpts parts of canto 2, informing readers that these passages describe "an island being gradually formed by the coral insects."[32] An 1854 obituary in New York's *The Eclectic Magazine* remembers Montgomery for his portrayal of "coral insects," and an 1861 geography textbook published in Philadelphia defines reefs as "the production of the coral insect," citing lines from canto 2 as an apt illustration of the process.[33] By the 1870s Montgomery's US status as a poet of coral insect labor was taken for granted so that when American poet William Cullen Bryant edited a popular anthology titled *A Library of Poetry and Song* (1871), he included a selection of passages from canto 2, titling the excerpt "The Coral Insect."[34]

The US transformation of parts of Montgomery's epic into some of the country's most beloved works of coral insect poetry provides a unique opportunity to learn precisely what kind of laborers Americans popularly imagined coral insects to be. More than "an act of labour," continual or collective, coral insect work was an all-consuming, intergenerational, extractive process that the workers could not escape, but did not seem to mind.

Here is the beginning of Bryant's excerpt, titled "The Coral Insect. From 'The Pelican Island,'" as it appears in his popular poetry anthology:

> . . . Every one
> By instinct taught, performed its little task,
> To build its dwelling and its sepulchre,
> From its own essence exquisitely modelled;
> There breed, and die, and leave a progeny,
> Still multiplied beyond the reach of numbers,
> To frame new cells and tombs, then breed and die
> As all their ancestors had done,—and rest,
> Hermetically sealed, each in its shrine,
> A statue in this temple of oblivion!
> Millions of millions thus, from age to age,
> With simplest skill and toil unweariable,
> No moment and no movement unimproved,
> To swell the heightening, brightening, gradual mound,
> By marvellous structure climbing towards the day.[35]

In this excerpt the creatures identified by Americans as coral insects indeed labor collectively and continually for a common end, much as Marx's reef makers in *Capital*: "every one" of the "millions" of organisms together "build" and "frame" the reef by contributing "skill" and "toil" to the "marvellous structure," which gradually increases to become a coral island.

Yet the lines also suggest a less lively prospect. Coral insects produce the reef by instinctive "toil unweariable," a term indicating the exhaustless potential of their labor. Each individual instinctively contributes not only their labor but also their material body in full, for the reef is "from [their] own essence exquisitely modelled." Thus it cannot be said that coral insects live and work *on* the reef; they *are* the reef, body inseparable from labor product across generations, "hermetically sealed" therein, permanently fused to "their ancestors" while continually breeding and dying. In return for their labor, generations of workers receive "dwelling" and "sepulchre," yet other life forms receive land to live on—for we last see the reef "climbing towards the day" to sustain lives above water while the builders are fully forgotten. Though each becomes "a statue," the reef is a "temple of oblivion."

A similar depiction of coral insect labor emerges from a different passage in canto 2 when it appears as an epigraph to a US periodical essay, "A Paper on Corals" (1848), which discusses the role of "coral insects" in reef formation:

> I saw the living pile ascend,
> The mausoleum of its architects,
> Still dying upwards as their labors closed—
> Slime the material, but the slime was turned
> To adamant, by their petrific touch.
> Frail were their frames, ephemeral their lives,
> Their masonry imperishable. All
> Life's needful functions, food, exertion, rest,
> By nice economy of Providence,
> Were overruled to carry on the process
> Which out of water brought forth solid rock.
> Atom by atom thus the burthen grew,
> Even like an infant in the womb, till Time
> Delivered Ocean of that monstrous birth—
> A Coral Island, stretching east and west.[36]

According to these lines, coral is made by the collective "labors" of tiny "architects," "dying upwards," "Atom by atom." As they work, "All / Life's needful functions, food, exertion, rest" are "overrruled"—meaning either controlled or set aside—by Providence, and the end result enables other life above the waves. The reef is a "masonry imperishable" for others to inhabit because it is a "mausoleum" for its builders. Their "frail" "frames" and "ephemeral" "lives" endlessly yield an enduring and ever-enlarging foundation that is not for them.

One reason that US readers so readily identified Montgomery's "architects" as "coral insects" must be their striking similarity to the reef-builders in Lydia Huntley Sigourney's "The Coral Insect," a poem that was first published one year earlier than *Pelican Island* and also became a transatlantic hit.[37] In Bryant's anthology Sigourney's poem appears alongside the identically titled excerpt from Montgomery's epic (above). Given that so few of my readers will be familiar with Sigourney's work, here is "The Coral Insect" in full, as printed in Sigourney's own anthology, *Poems* (1827):

Toil on! toil on! ye ephemeral train,
Who build in the tossing and treacherous main;
Toil on,—for the wisdom of man ye mock,
With your sand-based structures and domes of rock;
Your columns the fathomless fountains lave,
And your arches spring up to the crested wave;—
Ye're a puny race, thus to boldly rear
A fabric so vast, in a realm so drear.

Ye bind the deep with your secret zone,
The ocean is seal'd, and the surge a stone;
Fresh wreaths from the coral pavement spring,
Like the terraced pride of Assyria's king;
The turf looks green where the breakers roll'd,
O'er the whirlpool ripens the rind of gold;—
The sea-snatch'd isle is the home of men,
And mountains exult where the wave hath been.

But why do ye plant 'neath the billows dark
The wrecking reef for the gallant bark?—
There are snares enough in the tented field,
Mid the blossom'd sweets that the valleys yield;
There are serpents to coil, ere the flowers are up;
There's a poison-drop in man's purest cup,
There are foes that watch for his cradle breath,
And why need ye sow the floods with death?

With mouldering bones the deeps are white,
From the ice-clad pole to the tropicks bright;—
The mermaid hath twisted her fingers cold
With the mesh of the sea-boy's curls of gold,
And the gods of ocean have frown'd to see
The mariner's bed in their halls of glee;—
Hath earth no graves, that ye thus must spread
The boundless sea for the thronging dead?

Ye build,—ye build,—but ye enter not in,
Like the tribes whom the desert devour'd in their sin;
From the land of promise ye fade and die,
Ere its verdure gleams forth on your weary eye;—
As the kings of the cloud-crown'd pyramid,
Their noteless bones in oblivion hid;
Ye sleep unmark'd 'mid the desolate main,
While the wonder and pride of your works remain.[38]

Just as in popular excerpts from Montgomery's poem, here reefs appear as nature's celebration of communal labor for common good, sustaining the laborers as they produce land for humankind—even as some reefs cause shipwreck—while the details of reef-making disclose another story of labor.

As Sigourney's insects "build," "rear," "bind," "plant," and "sow," from birth until death, their bodies fuse together to make a foundation from which they are fully excluded and forgotten. They "enter not in" but rather "fade" into the "desolate main" "and die." The rhythmic "Toil on! toil on!" that begins the poem echoes in the closing stanza's "Ye build,—ye build," thereby describing, and even possibly cheering on, the ongoingness of all-consuming reef labor. And Sigourney's poem particularly emphasizes the lack of recognition that the workers receive. While their toil is carried out in the same "oblivion" that surrounds Montgomery's coral-builders (Sigourney, line 38; Montgomery, line 10), Sigourney's speaker stresses their absence from sight and memory by calling their labor "secret," "noteless," and "unmark'd"—the latter terms meaning "unnoticed" and "unobserved."

As the coral insect poems of Sigourney and Montgomery circulated through US print culture, they probably appealed to different US audiences, since Sigourney's is a lyric by an American woman renowned for domestic themes, while Montgomery's is an epic about evolution by a staunchly evangelical Scottish man. The eager consumption of both texts thus indicates that Americans with different cultural identities and interests embraced a common definition of coral insect labor as far more than collective labor (as Dana suggests) and as quite other than uplifting labor that brings social purpose, importance, recognition, liberty, or belonging (as writers from Margaret Fuller to Karl Marx and many British Victorian authors suggest). To countless Americans coral insect labor is work that wholly defines and consumes the laborers, while largely erasing them from the historical record, as the reef emerges from their bodies to become a foundation that sustains others and looks nothing at all like the congealed mass of generations that it comprises.

In these various reflections on coral, reefs give material form and familiarity, alongside validation through repetition, to a logic that is strikingly continuous with one that powerfully rationalized slavery: some forms of life are "confined by nature to bodily existences," and thus suited to be the objects (and never the subjects) of the history they propel.[39] Though slavery is nowhere explicitly named in any of these writings on reefs, its logic of differentiation and devaluation is everywhere visible in the "obdurate materiality" of the laborers, whose combined durability and insensibility renders them suitable to create and extend the foundations that enable another, and a less fully corporeal, form of life to flourish.[40] Slavery requires the violent reduction of some beings into bodies alone, into so much "calculable matter," as Baucom reminds us, and that violence could not have transpired as successfully without the quotidian recitation of what Barrett identifies as an "elusive cultural logic" that makes "various groups visible and divisible in the first place."[41] By imagining polyps as tiny, durable, insensible, laboring bodies endlessly forging the ground for another and higher life form, Americans recited that specific logic through a uniquely passive and peaceful reef metaphor, naturalizing an inherently extractive labor relation.

Between the 1830s and 1860s US authors working in multiple genres, including poetry, periodical essay, and short fiction, drew on the language and imagery of Sigourney's "The Coral Insect" and Montgomery's *Pelican Island* to produce new accounts of reefs, coral islands, coral insects, and even coral objects. Upon first consideration, these texts seem to depart from the source material in that they draw from reefs a specific lesson that the humblest forms of life are most worthy of value and attention because they sometimes produce the most important and lasting works.[42] These texts overtly call on readers to recognize, celebrate, and value essential workers—both human and nonhuman—who are the most constitutive yet least likely to be included politically, socially, and historically.[43] However, what appears as a celebration of the contributions of the most essential laborers always also celebrates the extractive conditions that consume them. For the call to recognize these laborers is repeatedly complicated by descriptions of their bodies that resonate with Sigourney's and Montgomery's descriptions, often amplifying and adding details that reinforce the fantasy of a fully corporeal and inexhaustible worker, alongside the absolute centrality of life-consuming labor to the strongest aggregate, which one writer identifies as the US.

## EXHAUSTLESS LABORERS

In 1835 the *Boston Pearl* published an anonymous poem called "The Coral Insect" that encouraged readers to acknowledge the importance of the least visible and most essential workers.[44] The poem's form dramatizes that lesson by staging a shift from "awe-struck" admiration of four "majestic" humanmade structures—the pyramids of Egypt, the Tower of Babel, the city of Rome, and the great wall of China—to a deeper "Reverence" that "awakes" when we turn our "gaze" from "man and human acts" to "the insect world."[45] What follows is an extended description of coral insects at work. Like the figures described by Montgomery and Sigourney, these creatures are "small as the sand" yet "more than giant's might" in their collective action. Through "ceaseless work," "race after race" of these beings "calmly" "prepare" their "tiny homes," which are also the foundations of "vast mountains," "hills," and "vales" for "man" to inhabit one day.[46]

With these indispensable laboring bodies in view, the speaker asks two rhetorical questions:

> Could human hands have strewn
> Piles of eternal stone
> And broad foundations in unfathomed seas?
> Could man so calmly urge
> Up through the boiling surge
> Towers of unmeasured height, and buildings such as these?

The poem concludes by instructing readers to "look at the meanest one, / And see thyself outdone; / The insect of the waves is mightier far than thou." While delivering that lesson in humility, however, the poem also makes another point: the most "awe-inspiring," "broad," and "eternal" structure imaginable is the very ground that we stand on, which emerges only through the "ceaseless," intergenerational, and all-consuming work of life forms whose most remarkable trait is an inexhaustible labor capacity: they "calmly" produce what is far beyond "human hands" to create, yet necessary for human life.[47]

A version of this message emerges in numerous imaginative writings on coral during the mid-nineteenth century. "The Coral" (1845), an anonymous poem in a New York periodical, asks readers to check their pride by looking beneath the waves, for "how mean would the grandest works compare, / That pride of man can form, / With the mighty power in progress there, / The skill of the insect worm!"[48] The poem also describes these insects in terms that would have been familiar to its readers: deep in "cerulean gloom" "unnumbered myriad swarms" form their own "shell," "lasting shroud" and "home and tomb," which covers over them while producing "spreading climes that yet will rise" for humans. As much as the poem positions reefs in their grandeur as reminders of human insignificance, it also indulges the fantasy of an essential laborer whose body endlessly yields a "wondrous" foundation for others by sheltering, entombing, and erasing itself.[49]

Many writings on reefs encouraged profound and overdue gratitude toward coral insects by suggesting that there would be no land at all without them. "Thou wondrous insect!" begins "The Coral Worm" (1836), a poem by one Mrs. Thompson in a Rochester periodical for women.[50] Like Sigourney's poem, this one apostrophizes the creature, and in similar language, for the animal "labors" "far down" in "silence," a "tiny architect" building its "fathomless abode." That labor is the noblest service imaginable, for it "buildest land" by making solid ground from sea, a labor that is both inexhaustible—for the animal will work until *there is no more sea*" states the speaker, quoting Revelations— and not particularly taxing for the worker, for from coral insect bodies "verdant islands rise, / As if enchantment with her magic spell / Spake into life an earthly paradise, / Where late was seen the liquid mountain's swell."[51] Joining Mrs. Thompson's speaker in awe at the land-building powers of coral insects was the narrator of an 1840 sketch by Edgar Allan Poe who deemed the coral insect "the architect of continents," and the character of Pip in Melville's *Moby-Dick* (1851) who sees "the multitudinous, God-omnipresent, coral insects, that out of the firmament of waters heaved the colossal orbs."[52] Some writers even wondered whether coral insects would shape the course of empire by creating new islands for European or American settlement.[53]

While coral insects drive human history by creating the material foundations of all land past, present, and future, for all their life-consuming and "exhaustless labours," they experience little or no physical duress, many writers assured their readers.[54] *The Works of Creation Illustrated* (185-), an evangelical natural history adapted from an English source and published in Philadelphia, praises coral insects as "tiny workmen" responsible for many Pacific Islands.[55] The author then allays any potential concerns about worker welfare, for "they appear perfectly contented with their allotted station, the food appointed affords them perfect satisfaction, and their internal organization is such that they do not even suffer from abstinence. Neither are they affected by the variety of the seasons, but, living in one temperature, appear almost exempt from disease, and generally enjoy a long-protracted existence."[56] In a different evangelical natural history, one B. F. Gilman of New Haven, Connecticut, excerpts passages from a number of popular writings on coral to explain the unique nature of coral insect bodies, including a history of Scotland that lauds these animals for their capacity to "construct islands and continents for the habitation of man" through "invisible, insensible toil."[57] The creatures are ever "working unseen, unheard" to "[extend] the dominions of man—who sees it not, and knows it not," and thus could easily have "despised" the laborers as "inadequate." Presumably to affirm these claims, Gilman next includes the popular excerpt from *Pelican Island* in which "millions of millions thus from age to age / With simplest skill and toil unweariable" build an island.[58]

Of all the period's reflections on coral indebted to the work of Montgomery and Sigourney, however, none as vividly describes exhaustless coral insect bodies or as overtly suggests the continuity between reef-making and US founding and expansion as "Sea-Life" (1866). A short story by Massachusetts author and educator Jane Andrews, "Sea-Life" celebrates the creation of Coraltown. This fictional Caribbean atoll was founded thousands of years ago by "emigrant" coral insects "looking for a new home," just as were "the Pilgrims at Plymouth."[59] Generations of these "little settlers" subsequently developed Coraltown into a full-blown island that preserves the lives of hundreds of passengers whose steamship had foundered in the Caribbean Sea en route from New York to California in 1865.[60] Thereby the coral insects of Coraltown teach humans that even the most seemingly insignificant individuals may produce massive and mutually beneficial results by choosing to work together over time; or, as the narrator explains, "people sometimes live in communities and divide the work as suits their fancy" and thus "the good Father teaches all his creatures to help each other."[61]

Yet this narrative of voluntary work for mutual gain erodes when Andrews turns from the wonder of coral islands rising from the sea to a fuller account of what happens under the surface. At Coraltown, an unsettling fusion of laborer and product prevails. "How do you like this little circular town?" asks the narrator, who then directs readers to notice that the atoll emerges because each polyp "from day to day fastens himself more and more firmly to the rock where he first stuck" so that "the part of his body touching the rock hardens into stone, and, as the months and years go by, the sides of his body too turn to stone, and yet he is still alive."[62] Coraltown, we learn, is made of the "stone bodies" of polyp "settlers" and "emigrants"—of generations of "fathers and mothers, brothers and sisters, uncles and aunts," of "children," of "great-grandfathers and great-grandmothers," into whom "their descendants" merge endlessly.[63] One day a speaking starfish visits Coraltown and asks the polyps to come out of their "houses," by which he means the reef. The polyps answer collectively that they cannot, for "these [walls] are not our houses, but *ourselves*."[64]

This account of Coraltown sits rather uneasily alongside the story's explicit fable of voluntary labor for collective gain. The production of coral, as Andrews describes it, forecloses all prospect of the producer's very existence, bodily or otherwise, apart from the material they produce. Reefs are countless generations of laborers, solidified, with a disproportionate benefit accruing to others, in this case the hundreds of human passengers aboard "the great steamship Golden Rule," for whom Coraltown is a "harbor of refuge" that sustains them until they are rescued and resume their trip to California.[65]

What "Sea-Life" finally tells us is just how easily a bright island of refuge for all can require and conceal past and ongoing coercions and exclusions of its most constitutive workers. It also confirms that coral transmits this lesson, even despite authorial intent, and sometimes in direct defiance of it. And it suggests that this is the real story of labor in the US by describing the origins of Coraltown in terms that evoke the Puritan origin story of America as a "city on a hill": Coraltown began as "a little hill" settled by "emigrants looking for a new home . . . just as the landing of the Pilgrims at Plymouth was the beginning of Massachusetts."[66] The story of American exceptionalism takes a strange turn at this Caribbean atoll, however. As Coraltown models the US's origin and expansion in collective self-sustenance, it also indicates our past and ongoing indebtedness to enforced labor, thereby reminding us of the Indigenous and Black persons whose active and skilled contributions rendered them inseparable, yet also excluded, from the country that they built.

## "THE CORAL PEOPLE"

None of the texts on coral and labor thus far examined explicitly compares coral insects to any particular group of human laborers in the nineteenth-century US. Many of these texts encourage readers to compare coral insects to human laborers generally, sometimes by describing the insects in terms that could mean a group of either human or nonhuman beings: Andrews calls them "settlers" and "emigrants," and Sigourney a "train." Taking the association one step further, the narrator of a fairy tale by Louisa May Alcott calls coral insects "the coral people."[67] But if, as I have been arguing all along, reefs encoded an underexamined dimension of the US political imagination of *human* labor under the combined rise of slavery and industrial capitalism, then it would be important to know who exactly "the coral people" were.

According to a number of white, male, non-working-class writers, the human equivalents of coral insects, as they have been thus far described, were white working-class laborers who fulfilled a variety of essential roles demanding physical exertion. Some authors made this claim in the course of merely defining coral. "You have heard a great deal about the wonders performed by little animals called coral insects," begins "A Talk About Coral" (1856), an essay that appeared in several US children's periodicals during 1856.[68] Reefs, readers of *Woodworth's Youth's Cabinet* learned, are made by creatures "pretty low in the scale of animal life," and who "cannot have much sensation," or truly "any at all," particularly given that they can be cut into pieces, with each piece then quickly becoming "a complete animal in itself, and going straight to work."[69] The essay proceeds as follows:

> The coral polypes are the *stone-masons* of their race. They build, for the most part, of chalk. They do not walk about much, but are accustomed to spend their days in one place. These animals are great scavengers. Do you know what a scavenger is? Perhaps you have seen men sweep dust, and dirt, and straws, away from the streets of a city; such men are called scavengers, and we are greatly obliged to them for the work they perform, since decayed animal and vegetable substances, when they become putrid, injure the air we breathe. Impure substances also injure water; therefore, to assist in cleansing the seas, the Lord God has seen good to make the polypiferous family the great scavengers of the ocean; and a very numerous and happy set of workmen they are, ever delighting in their appointed labors. They have no brooms to sweep with, no carts to hurry away what they collect; but every bad morsel of decaying substance that floats past them they fold in their arms, and with joy they pour the offensive bits of refuse into their living tubes, where the juice in their bodies turns what they take into part of their own living substance.[70]

In the course of casually "talking" about coral with America's youth, this essay encourages children to imagine human workers, both stonemasons and street cleaners, as part of a lower life form, biologically adapted to "spend their days in one place" performing repetitive, confining, all-consuming labor that sustains other people. For, the author suggests, polyps are best imagined as a "very happy set of workmen" who thrive by doing nothing other than building a structure for others, just as human stonemasons do, and by cheerfully consuming waste that

they transform into their bodies and our livable environs, just as human scaven-
gers do. The idea seems to be that city buildings, clean streets, and breathable
air require a class of laborers whose bodies differ from "ours." The author repeats
familiar assurances that coral insects are perfectly adapted to their work, "ever
delighting in their appointed labors," and feel no pain—they work "with joy"—
while seamlessly transferring these qualities to white, working-class humans.

That comparison helps a different writer define coral for adults in "A Day on
a Coral Reef," an essay that appeared in two US religious and literary journals
during 1864.[71] Part travel narrative and part natural history, the essay is about
"curio-hunting" along a coral reef in the Red Sea where the speaker and a friend
spend a "splendid" day filling their jars, pockets, and bags with as many mol-
lusks, star-fish, shells, corals, and other marine specimens as they can carry,
pausing only for a picnic lunch.[72] In the course of their excursion the speaker
wonders, "What are corals?" and "What are the coral insects?"[73] He answers by
dispelling the "common idea that coral lives in a cell that it has built, and may
be seen looking out of the top of it, very much like a chimney sweep rejoicing
in his strength at the top of a chimney."[74] Rather, "the truer image would be
the head and arms of a sweep, but nothing more of him except his skin tightly
stretched over a blocked-up chimney. The bricks should be inside him, for the
hard parts of the corals are certainly inside them."[75] According to the author,
coral polyps are like chimney sweeps, if the latter could somehow merge their
bodies with and consume the chimney materials so that they would have literally no bodily existence or needs apart from their work. While the author
differentiates coral polyps from actual chimney sweeps, the comparison is only
possible and effective because of how easily certain white laboring bodies can
be imagined to serve extractive undertakings; it is not that hard, in other words,
to suppose that the people sweeping the chimneys are more naturally suited to
and fulfilled by bodily labor than the people living in the houses with chimneys.

During the 1850s the comparison of coral to white working-class laborers
jumped from natural history to more overtly political discourse when white
writers and speakers evoked it with the intention of highlighting the overlooked contributions of the white working classes. According to "The Greatness of Little Things" (1855), a short essay widely reprinted in newspapers and
periodicals after its first appearance in Philadelphia's *Public Ledger*, a coral reef
reveals that "the smallest tenants of the ocean, not the largest, are its most
important occupants"; just so, those persons whom "historians, statesmen
and politicians" overlook are the true "monarchs of this great and wide sea of
politics."[76] For "the people, the industrious classes," "the poor," "the masses"
are the ones who "form the strata by which the geologists of history will here-
after mark the progress of this age."[77]

While demanding education reform for "the people," however, the author also normalizes their extractive relation to the non–laboring classes. For "the industrious classes," he explains, are those who "form the strata," just like "the coral insects" in "Lieut. Maury's recent work."[78] The work in question was a recently published natural history of the ocean, *The Physical Geography of the Sea* (1855) by American naval officer and oceanographer Matthew Fontaine Maury. According to Maury, the coral islands, reefs, and atolls of the Pacific Ocean were "built up of materials which a certain kind of insect quarried from" ocean currents that served the insect as "*hod carriers*," an obsolete term for humans who "carry mortar, and sometimes bricks and stones, to supply builders at work."[79] Filtering through the animal's body, those raw building materials fully sustain the coral insect, or "little mason," and thus we see God's providence: coral insects work "apparently for their own purposes only," while generating the essential foundation for others.[80] By way of Maury's coral insect simile, then, the author of "Greatness" indicates that white workers generally are "the strata," their bodies the ground that others stand on and forget.

A final instance of this comparison more sharply highlights its capacity to elide African slavery from the US political imaginary of labor under industrial capitalism. When speaking before the Missouri House of Representatives in February 1857, Benjamin Gratz Brown, representative from Saint Louis, closed his passionate argument for the abolition of slavery in Missouri by arguing that one "must go below the shining surface of society," to find "the true builders."[81] For while "we read in the voyages of mariners, stories of enchanted isles," it is only when we "fathom the depths" that we see these isles "rest upon vast reefs of coral, whose structure has been the work of millions of infinitesimal-animalculae laboring for ages and ages, with unending toil"—not Black enslaved persons but rather "the laboring classes struggling on in the daily avocations of life," those whom the end of chattel slavery would most benefit by opening up new jobs for white workers to fill.[82] For "upon the wharves, and along the streets, in counting-house, and workshop, and foundry, wherever labor was to be performed, their hands have been in ceaseless employ, and to them should be awarded the praise of all that has been accomplished."[83] According to Brown, if wage labor has not lived up to the appellation of "free labor," then this is only because white laborers have been hampered by having to compete with enslaved laborers, whereas the end of slavery would mean "emancipation" for the country's true "coral insects," "the white race."[84]

In reality, it was enslaved persons of African descent who could never enter alive the polity whose foundations they produced so laboriously across generations and whose conditions of coercion were most widely rationalized by a biological account of life that shares a number of features with that of coral insect life, as popularly imagined. Yet in all the research I have conducted for this book across a decade, I have come across only a single, fleeting instance of a white US writer comparing coral insects to enslaved Black persons.[85] That silence is instructive in several ways. It tells us, first, that white non-laboring classes routinely and tacitly told one another that there needed to be one class of people to "form the strata," "form" meaning both "to shape" and "to be the material of."[86] One group had to produce, with their labors and bodies in full, "the basis" on which others "will develop"[87] The stronger and more enduring and expansive the "structure," the more living bodies consumed wholesale, yet also wholly fulfilled by their work since their "internal organization" is such that they have very little sensation, or none at all. And as widely as this script of labor circulated, as continuous as the imagined position of the coral insect became with the political condition of enslaved persons, contemporaries did not openly mark the resemblance.

By calling polyps "coral insects," then, US writers did not, by and large, betray their ignorance of science. Instead, they preserved the power of coral to speak to human concerns. We might productively understand the popular insistence that polyps are insects within a long literary and philosophical tradition that elevated insects to beings worthy of careful attention, and even affection, for what their behavior could reveal about the structure of human societies and the role of labor within those societies. As insects, polyps were on par with other "social organisms," such as the bees and ants that historian A. J. Lustig describes as valid, useful, and relevant repositories of human political imagining.[88] Bees in particular have served as a popular political allegory in philosophical and literary works at least since Bernard Mandeville's *Fable of the Bees* (1714); bees were also frequently imagined as colonial, collective creatures, whose individual "labors" produced a marvelous structure. A popular idea, for example, was that a hive flourishes, sustaining both bees and those who consume the honey, merely because each bee pursues its own needs and desires. Yet there is this crucial difference between imaginings of hive and reef: bees do not die as they labor to produce a structure that rises ever upward from their bodies to support those who never contributed to it. Coral insects, by contrast, modeled a reality that many Americans found notoriously difficult to openly state: a country formally dedicated to the ideal of individual liberty for all had always required the disproportionate labor of those it excluded.

This history of coral insects in US literature and culture establishes the power of coral to help humans conceptualize the ideologies of labor that capitalism deploys and relies on, even as it frequently conceals them. For even when these writers employ coral to call on their readers to perceive and acknowledge the constitutive role of laborers, or even to fund education of these laborers, the laborers themselves appear in this system exclusively as bodies, a condition that facilitates their living entombment and oblivion. Or worse, some laborers do not appear at all, since the formative and sustaining role of Black persons is fully elided when white writers identify white wage laborers alone as the polity's coral insects. That all of this transpires through the figure of a uniquely harmonious reef metaphor finally indicates how easily a system of labor appearing to offer social and economic justice can silently and violently extract the life and labor of the many for the benefit of the few.

How can it be, then, that as we will see in some of the following chapters, many white women, and many Black women and men, willingly self-identified as coral insects? As the next chapter will explain, these writers drew on a different conception of coral insects to develop a different political imaginary of labor. In that version coral insects are more than laboring bodies, and reflections on coral achieve more than a numbing routinization and familiarity with the ongoing realities of exploitative labor. Instead, coral abets the efforts of these writers to confront and challenge capitalism's requisite coercions and exclusions, its living entombment of the most constitutive laborers. Coral becomes a daily reminder and exposure of the ongoing violence of reducing persons to matter and an enjoinder to reform the reef—or make a new and different one. To preview that other and more hopeful meaning of coral that awaits the reader, let us consider a woman made of "korl."

INTERCHAPTER

# The Korl Woman

"Nothing remains to tell that the poor Welsh puddler once lived, but this figure of the mill-woman cut in korl," reflects the narrator of Rebecca Harding Davis's "Life in the Iron-Mills; Or, The Korl Woman" (1861).[1] A stark exposure of industrial capitalism's exploitation of human workers, the story brings readers into the lives of Deborah and Hugh Wolfe, confronting us with the exhaustion, hunger, and "incessant labor" that consumes countless "thousands" of immigrant mill workers like them.[2] Central to the narrative is a sculpture that Hugh carves out of "the refuse from the ore after the pig-metal is run," a substance that Davis calls "korl."[3]

"The korl woman" provokes the story's most pressing questions—"Is that all of their lives?" and "Who is responsible?"—and then outlasts her maker to stand in the narrator's library and mutely prompt the recriminating question, "Is this the End?"[4] Industrial labor, Davis means to say, in its wholesale reduction of some persons to laboring bodies, produces "a reality of soul-starvation, of living death" for the laborer, while enriching others, such as the overseer and the mill-owner who refuse accountability for the laborer's plight.[5]

Davis's "korl woman" has generated much scholarly analysis and debate, yet no one seems to be able to explain exactly what "korl" is. The word is unique to Davis's story, and the narrator's definition of it—"the refuse" from iron extracted from ore—has led one scholar to suggest that "korl" may be a colloquialism for *scoria*, "the slag or refuse left after metal has been smelted from ore."[6] Yet scoria, or slag, is black, or at least dark-tinted, possibly gray at times. Davis's "korl," by contrast, is remarkable for its "light . . . flesh-colored tinge" and is sometimes described as "white."[7] Possibly building on such descriptions, another scholar points out that "korl" is "a homonym of 'coral'," and thus complements the "water imagery" that fills the story.[8] Could "korl" be coral?

Korl is what remains, both after the laborious process of iron extraction and long after the life of the laborer, to remind the narrator ensconced in her library of the constraints and exclusions that industrial labor requires. It stands, "a working woman's body . . . strong, tired, and dissatisfied," neither quite alive nor dead, according to those who observe it, and always refusing romance and sentiment, as Jean Pfaelzer observes.[9] At the start of the story the narrator tells us that this is "an old story," and "you may think it a tiresome story enough," one among "thousands of dull lives like its own . . . massed, vile, slimy lives," but "I want to make it a real thing to you."[10] To make the exploitation real, Davis seems to say, what better material than coral, which refuses to let us look away from the human costs of capitalist labor and always demands "redress"?[11]

Whatever Davis's "mill-woman cut in korl" is made of, she pierces the political fantasy of an inexhaustible laboring body, and provokes questions about the US's dependence on extractive labor.[12] By prompting observers to ask "Is that all of their lives?" and "Who is responsible?," the korl woman generates the same questions about labor that, as I will show in the following chapters, coral polyps and coral reefs encouraged many other US writers to ask and answer; these questions and answers tell us that the US political imaginary of coral took directions other than the one explored in the preceding chapter.

Davis's work shares a sharp critique of wage labor with an earlier and less widely read work, a poem about coral by one "Anna" that appeared in *The Operatives' Magazine*, a short-lived journal edited by and for women working in a different kind of mill, the textile mills of Lowell, Massachusetts.[13] Anna's poem, "The Coral Insect" (1842), identifies these women—the "mill girls," as they were often called—as the polity's coral insects. Especially because this remarkable poem has not been reprinted, it is worth quoting in full:

Down in the deep and silvery sea,
An insect built his coral bower;
No jewel'd prince, more gay than he,
In splendid hall, or regal tower.

And many a pearl that glistening there,
The sea-grass tall hung waving o'er;

And curious shells and diamonds rare,
Lay scattered on the shining floor.

But from the deep the stranger tore
The coral bower, with eager hands,
And to his home the prize he bore,
Far, far away in distant lands.

And many on the treasure gazed,
Wrought with such loveliness and care,
But while the sea-plant many prais'd,
None deemed an insect wrought it there.

Thus genius with its unseen might
Wreathes garlands for the rich and great:
Wakes dreams of beauty, wondrous bright,
While insect-like meets "lonely fate."[14]

The poem subtly analogizes coral insects to human weavers: just as coral insects "wrought" the reef from which coral specimens ("treasures") are torn to adorn the curiosity cabinets of "strangers," "insect-like" beings "wreathe"—a word that evokes weaving or spinning—"garlands" for "the rich and great."[15] And just as most people tend to "gaze" upon and praise the "loveliness" of coral "treasures" without crediting the "insect" who "wrought" these items, so "the rich and great" look upon their "garlands" and dream "dreams of beauty, wondrous bright" without giving a thought to those "unseen" and "lonely" weavers who produced the material that enables their dreams. The extractive, even rapacious, relation between the coral insects who weave the reef and the "strangers" who tear away pieces of "the coral bower, with eager hands," merely for their luxury, mirrors that between the human weavers of cloth and "the rich and great."

Yet what sets Anna's poem apart from the many texts by white non-working-class men that analogize coral insects to white working-class laborers—and what aligns Anna's work instead with Davis's text—is the obvious critique of these conditions. Like Davis's woman made of korl, the coral specimen in Anna's poem provokes readers to acknowledge that the workers are not mere bodies with no existence apart from their labor. Here coral prompts Anna to say that the young women endlessly spinning and weaving in silence, transforming raw cotton into rich garlands for a nation that neither sees nor remembers them, are also endowed with "genius," evident in the wonder that their creations provoke. The tragedy is that their "genius," as it extends into the world beyond their bodies, does not bring them recognition, gratitude, or material support. Instead, that genius becomes yet another inroad, alongside their physical laboring capacity, through which to exploit them.

By lamenting this condition and hoping that it could be otherwise, Anna's poem, alongside Davis's korl woman, tells us that in the US the political meaning of coral itself changed across time and context and depending on whom one asked. This is so because worldwide circulations of knowledge about coral from many centuries and cultures made different meanings of coral available, and appealing, to different groups of nineteenth-century Americans.

# 3
# Fathomless Forms of Life

The previous chapter established that many Americans repeatedly (and delightedly) imagined coral insects as fully corporeal forms of life dedicated to an all-consuming, intergenerational, extractive labor process that they could not escape yet did not seem to mind—or that they were even fulfilled by—as they produced an ever-expanding reef that erased them while enabling others to flourish. That understanding of coral insects upheld the claim that a collective thrives when each individual labors for their own gain. It elided exploitative labor and laborers. And it did so, moreover, at a historical moment when the coercive labor of multiple human groups increasingly sustained and expanded a country that celebrated its dedication to unprecedented human liberty. Why, then, would so many marginalized persons living in this society fervently self-identify as coral insects and encourage themselves and one another to be like coral, to build a reef?

The list of brilliant and differently marginalized US writers and activists who did so, from the widely known to the obscure, is quite stunning. It includes numerous white women, such as abolitionist and best-selling novelist Harriet Beecher Stowe (1811–96), historian and translator Elizabeth Wormeley Latimer (1822–1904), early feminist and philosopher Ellen M. Mitchell (1838–1920), and Boston-based adoption reformers and sisters Anstrice Fellows (1800–75) and Eunice C. Fellows (1802–83). It also includes a large number of Black men and women, such as abolitionist and author James McCune Smith (1813–65), writer and civil rights advocate Frances Ellen Watkins Harper (1825–1911), abolitionist and journalist Jonas Holland Townsend (1820–72), attorney and journalist Ferdinand Lee Barnett (1859–1936), African American minister Abraham Lincoln DeMond (1867–1936), and Cleveland-based civil rights leader Frank Lee (ca. 1842–1907).

I will analyze the words or work of each of these people in the next chapter. As the necessary basis for that analysis, this chapter answers the question, What are the grounds of their relation to coral and of their specific identification with coral insects? For those grounds simply cannot be the cultural understanding of coral reefs and coral insects that took shape within the writings examined at length in the previous chapter.

Two other intersecting domains of knowledge about coral also infused popular nineteenth-century US conceptions of it. Together these domains produced a different frame of reference through which differently marginalized groups identified with coral and drew on the nature of reefs to give voice, form, and power to their own, anticolonial political visions, which always involve the claim that the most robust polities expand by sustaining, rather than displacing, a vast variety of others.

One of these domains of knowledge, which I examine in the first half of this chapter, is coral's long and storied European natural history from the classical period through the Enlightenment. My focus will be on a particular episode that fascinated nineteenth-century Americans: the eighteenth-century French naturalist Jean-André Peyssonnel's claim that reefs are made by the "coral insect"—the first documented use of that term—in his *Traité du Corail* (1726).[1] Within the annals of Enlightenment coral science, this moment eventually led to the classification of coral as an animal, thereby putting an end to the centuries-long debate over coral's taxonomic place.[2] Yet what mattered to nineteenth-century Americans looking back at this moment was not that naturalists *did* eventually place coral firmly within the animal kingdom but rather that they had failed to do so for centuries beforehand.

At stake for Americans in this watershed scientific moment and the category crisis leading to it was the question of whether external features could index the "real nature" of beings, nonhuman or human. To many Americans, this question was also political: according to a nineteenth-century biological account of race that emerged to rationalize slavery and other colonial projects, race and gender were perceptible bodily features that enabled people to measure what was harder to see—intelligence, character, creative capacities—and thereby to rank persons as more or less suited to bodily labor.[3] That determinist logic shaped nearly all facets of US culture, so that in 1844 Ralph Waldo Emerson diagnosed the century's "impudent knowingness," its certainty that "the law of [another's] being" was legible by "such cheap sign-boards" as color and shape.[4] The natural history of coral, however, cast doubt on the correspondence between visible and invisible. And though this doubt did not deter scientific racism or the labor regimes it rationalized, it did sometimes check the "impudent knowingness."[5] As coral's historic taxonomic contest reemerged and circulated alongside coral itself in nineteenth-century US schoolrooms, boarding houses, churches, and parlors, it encouraged generations of Americans to speculate that knowledge required resisting the impulse to name and know and rank based on appearances.[6]

Simultaneously there arose yet another domain of coral knowledge, to which I will turn in the second half of this chapter: the nineteenth-century science of coral reefs, and particularly Darwin's account of reef formation in *The Structure and Distribution of Coral Reefs* (1842), a text widely excerpted and discussed in the US. My focus will be on several of Darwin's key insights that American writers and activists, white and Black, took up and transformed into key political claims and practices. Of importance to these Americans was Darwin's explanation that reef growth is unpredictable because the smallest movement of a single entity could dramatically alter the entire structure over time.

Perceived through these alternate frames of reference, coral insects were not laborers consumed and fulfilled and forgotten by building the structure; they were fathomless forms of life. As figures of hidden complexity, warnings against classification, and models of the power of indeterminate, relational emergence, reefs now became an appealing political metaphor to persons marginalized by race and gender who drew on coral to imagine that a different political future could be designed and created by *persons* who had been variously underestimated, dispossessed, exploited, or forgotten, despite their absolute centrality to the US polity. Reefs even became a natural challenge to capitalist individualism since they grew not because individuals labor to sustain themselves, but rather because they forged relations, past and present, that might transform the structure.

Today we still say that it is astonishing to look at a piece of coral, whether as heirloom jewelry or as museum specimen, and remember that it is not a stone or a plant but rather the remnants of an animal body that is more complex and unpredictable than it tends to appear. Just as astonishing is to remember that this body emerged through extraordinarily complex relations with other animals and with plants and the sea. In some ways, then, this chapter is a prehistory of that experience of coral and a plea to understand it as a vital chance to practice looking closely, while knowing one cannot trust one's senses to discern and evaluate and predict—a politically vital lesson from nonliving coral that is possible only because of what living coral teaches us, and one that we have yet to enact widely enough.

## Hidden Complexities: Peyssonnel's "Coral Insect" (1726)

Today the eighteenth-century French naturalist Jean-André Peyssonnel is all but forgotten by most Americans. People no longer sing about his contributions to coral taxonomy. Books are no longer dedicated to Peyssonnel. Schoolchildren do not recite lines from his treatise, *Traité du Corail* (1726) or tell of the trials that Peyssonnel suffered as the result of knowing, long before anyone else did, that coral is a more complex and surprising form of life than it appears to be. But during the nineteenth century, Americans did all of these things as they told and retold (and significantly revised) the now relatively neglected story of Peyssonnel's key contribution to the natural history of coral: his discovery of "the coral insect," which is what he deemed the tiny animal that he observed in the process of producing coral reefs. Since I imagine that most readers today will be less familiar with Peyssonnel's discovery, its aftermath, and its meaning within the larger history of coral taxonomy than nineteenth-century Americans were, I begin with that story.

Jean-André Peyssonnel (1694–1759) developed an early and lasting love of coral. Growing up in the Mediterranean port town of Marseille as the son of a renowned physician, Peyssonnel had ready access to the sea and to the men of science who passed through a city known for its medical school. When in 1706 Italian naturalist Luigi Ferdinando Marsigli traveled to Marseille to study ocean currents, he was hosted by the Peyssonnel family, and Jean-André, then aged twelve, watched as Marsigli studied the corals retrieved by Italian coral fishers who sailed annually to Marseille to harvest raw coral for the global coral trade centered in Italy.[7] What Marsigli saw of coral at Marseille persuaded him that coral was not a mineral—as was then widely believed—but rather a kind of flower and thus part of the vegetable kingdom. In *Histoire Physique de la Mer* (1725) Marsigli included color images that emphasize coral's vegetative, branching, plantlike features.[8]

By the 1720s, however, Peyssonnel had already begun to develop and test a different theory about the nature of coral. He traveled to sea with coral fishers, first at Marseille, and then along the "Barbary Coast," as the shore of North Africa was called, observing coral in its living state. After meticulous experimentation he concluded that coral is, in his words, the "work" of "the coral insect," which moves "its claws or legs" and thus is decidedly animal—but when he shared that discovery with the *Académie des Sciences* at Paris in 1726, ridicule and rejection were the official responses.[9] The "insects" that Peyssonnel observed might well be mere "visitors" to the coral plant, claimed prominent French naturalist René de Réaumur, while no less than Voltaire quipped that it would make as much sense to find a chunk of cheese with mites in it and thus to claim that the mites had made the cheese.[10] Coral remained in the plant kingdom, and the humiliated Peyssonnel, his finances and scientific reputation in ruins, went off to the West Indies for the next thirty years, during which he was all but forgotten in Europe as coral science moved on without him.[11]

The tide turned for Peyssonnel when, fifteen years later, in 1741, Abraham Trembley discovered the polyp, a free-floating creature that exhibits both plant

and animal characteristics and reproduces when cut in half.[12] Trembley's discovery opened new discussions about the boundaries between plant and animal
kingdoms. As naturalists began to accept that this profoundly plantlike being
was indeed an animal, they likewise began to suspect that Peyssonnel had been
right. Reef-producing animals were declared a type of polyp, a fact that even
Linnaeus, a late convert to the idea, accepted in 1758. Efforts to rehabilitate
Peyssonnel were made during his lifetime—apologies publicly issued, his discovery acknowledged in print, and his membership in the Royal Society granted
in 1752—but by then it was too late.[13] Peyssonnel's findings had already been
superseded by the work of other naturalists, most notably John Ellis, and apart
from a brief return to France, Peyssonnel remained in Guadeloupe where he died
in relative obscurity in 1759.[14] In the annals of Enlightenment natural history,
Trembley tends to receive credit for discovering the animal nature of coral, Ellis
for confirming it, and Linnaeus for (reluctantly) validating it.[15]

Among nineteenth-century Americans, however, the story was different:
Peyssonnel was the decided hero of coral's taxonomic drama, with Trembley
and Ellis relegated to second place at best. That version of the story began to
take shape in early American reference works that follow the lead of some of
the late-eighteenth-century European naturalists who had tried to restore Peyssonnel's reputation. The entry on coral in James Akin's *Encyclopaedia* (1790),
the earliest encyclopedia published in the US, describes Peyssonnel as "the
first who threw a proper light upon the nature and production of coral," though
lamentably "this ingenious naturalist" was "little regarded" until "Mr. Trembley's discovery of the fresh-water polyp."[16] A popular natural history follows
suit in 1799, explaining that the "nature and origin of coral" was "much disputed," and coral "ranked among . . . marine plants" until "M. de Peysonnel"
came along.[17] Samuel Miller's *Brief Retrospect of the Eighteenth Century* (1803)
declares that "*corals*, at the beginning of the eighteenth century, were reckoned
among the number of marine *plants* . . . until M. Peysonnelle . . . ascertained
their *animal* nature," which Trembley and Ellis only corroborated.[18] These and
other early US texts ensured that Americans got the story straight: though
"little regarded," the "ingenious" Peyssonnel had been right.

As the century proceeded, however, new details about Peyssonnel emerged
in popular and scientific US writing about coral's natural history. More than
a neglected genius, Peyssonnel was now an outsider who had risked everything to challenge the scientific establishment in the name of justice for forms
of life that had been ranked unfairly low in the great chain of being because of
how they happened to look. Prominent naturalists from the classical period to
the eighteenth century, though known for their great observational powers,
had been misled by coral's stone or vegetable form and were even oppressors

(according to some nineteenth-century sources) for denying coral its rightful place on the chain of being, while coral itself went unacknowledged for its extraordinary nature because it happened to look like a plant.

These sources collectively transformed the natural history of coral into a cautionary tale about the danger of placing too much faith in taxonomic classification, and in human abilities to discern "true nature"—both nonhuman and human—by external forms. That caution became central to the everyday meaning of coral in the nineteenth-century US.

In 1846 American zoologist James Dwight Dana introduced his treatise on zoophytes by framing Peyssonnel as unique among all natural historians of coral for not being taken in by coral's vegetable form. Corals are "forms of life" that individually resemble flowers in "external figure" and are collectively "so like the trees and shrubs of land vegetation, as to have deceived even the philosopher till near a century since," begins Dana.[19] In fact, "all the early authors," from Dioscorides (ca. 40–90 A.D.) to Marsigli, "arranged corals along with marine plants"—until, that is, Peyssonnel arrived on the scene and "ventured to combat the prevalent opinion."[20] For that venture Peyssonnel received decades of "derision" and dismissal, from "the distinguished Réaumur," who insisted in 1727 that "the vegetable nature of zoophytes [was] too well ascertained to be a subject of discussion," to the "umpire of science" Linnaeus, who was at first quite "unwilling" to believe, even in the 1750s.[21] In Dana's version of the story, an outsider "combats" the establishment, suffering because others were misled by coral's visible attributes.

Some of these sources more explicitly frame Peyssonnel as risking his reputation on behalf of an unjustly disregarded group, lending both bravery and morality to his capacity to see beyond coral's appearance. *Wonders of the Deep* (1869), a popular natural history of the sea by Maximilian Schele de Vere, deems corals "altogether a strange, mysterious race," their nature disputed for centuries as men quarreled "pertinaciously" over where they "belonged" in the scale of being—that is, until the start of the last century, when one man "raised [them] to the dignity of animals": "a young physician from Marseilles, called Peyssonel . . . first discovered their true nature."[22] When Peyssonnel informed Réaumur, however, "the illustrious naturalist was still so firmly bound by precedent and scholastic method that he refused to endorse the bold doctor's statement," and when Réaumur later tried to make recompense, it was too late: Peyssonnel "had gone, in disgust and despair, to the West Indies, and there he had disappeared from the sight of men, so that to this day we know neither the time nor the place of his death."[23] Schele de Vere's Peyssonnel is a "bold" "young" man at the margins who "raises" the coral "race" to "dignity," while the "illustrious" fail to perceive the "great value" of coral and Peyssonnel, who slinks off to the edges of empire in "disgust," "despair," and oblivion.[24]

That so many great men of science, known for their extraordinary observational powers, should have missed the true nature of coral is a fact that ought to make us wonder what we of the present are failing to perceive, even now, when we place too much faith in our own observational powers, and in taxonomy itself. This conclusion is, at any rate, what Schele de Vere draws from the story of Peyssonnel and coral. According to Schele de Vere,

> It seems now astonishing how men could quarrel so long and so pertinaciously over the apparently simple question, whether corals belonged to the vegetable or the animal kingdom. More fortunate in this respect than many other organic forms, whose social status is not yet recognized, corals were already, in the beginning of the last century, raised to the dignity of animals. This was not achieved, however, without much trouble and much ludicrous blundering.[25]

Since the phrase "social status" means "a *person's* position, standing, or relative importance in society," Schele de Vere's analogy of corals to "other organic forms, whose social status is not yet recognized" suggests that corals ought to teach humans not to devalue other humans based on their "organic forms."[26]

This interpretation of coral seems to be what made it so appealing to authors of evangelical tracts and books, "the most widespread form of popular media in the antebellum United States."[27] For a number of these authors, encounters with coral specimens and other objects provided an everyday opportunity to both reflect on the folly of evaluating physical features as predictive of nature, character, or capacities, and practice avoiding that mistake. While these authors do not always explicitly suggest that people should apply that way of seeing and evaluating to human groups, as Schele de Vere does, they nearly always hint that one value of coral to humans is its unique power to help us both understand and avoid a tendency to disregard whatever in nature appears lowly to us.

*The Wonders of the Deep* (1836), a slim evangelical natural history of the sea by Margaret Coxe that was printed by New York's General Protestant Episcopal Sunday School Union, tells the story of a family's summer vacation on the New England seashore where, one rainy evening, mother and children gathered in the parlor to peer at coral.[28] "I have, my dear children . . . brought you something to look at," says one Mrs. Stanley, removing the cover of a box and exhibiting "to the eager gaze of her children a variety of specimens of that wonderful insect 'the coral-worm.'"[29] Collected from the shores of New Jersey, the Mediterranean, and the West Indies, the coral specimens range in color from "a brilliant scarlet" to "the most snowy whiteness," some "branched like a delicate shrub" and others "shaped like an immense leaf."[30] Yet those various vegetable forms belie another and higher nature, Mrs. Stanley explains:

> It was, my dear children, a long while before it ever entered the minds of those who were interested in the study of nature, to conjecture that the beautiful specimens of coral with which the ocean abounds, were animal productions. So that in former days, among those, too, who had made them the subject of particular examination, there were some who ranked them among minerals; while others considered them as a branch of the vegetable creation. But they are now decidedly ascertained to belong to the animal world.[31]

What matters to Mrs. Stanley is that her children should know not only what coral is but how long coral was *misplaced*, even by those who had made corals "the subject of particular examination." Upon first seeing the specimens, the children themselves almost underestimate coral polyps by wondering how "a creature so diminutive" could make reefs.[32] But Mrs. Stanley explains that knowing coral means disregarding its appearance. For "we see that, insignificant as they appear to be in the scale of creation, they have succeeded in accomplishing what man with all his proud superiority has never dreamed of attempting."[33] The lesson is to avoid the mistake that "long" misled previous generations and kept many "persons ignorant of [coral's] real nature."[34] And while Coxe does not ask readers to apply this lesson to their evaluation of any particular group of humans, she did later write a two-volume history of women, *Claims of the Country on American Females* (1842), in which she frames her discussion of overlooked women in strikingly similar language. Women, Coxe writes near the start of volume 1, are more "deserving of attentive consideration," though they have been "greatly degraded in the scale of being."[35]

Another evangelical natural history hints more directly that more than coral taxonomy is at stake in rightly evaluating coral. *The Coral-Maker* (1844), a small square pamphlet printed by the Sunday School Union of the Methodist Episcopal Church, teaches readers to see coral through the lens of its historic category crisis. The "resemblance" between corals and plants meant that corals "were long considered as within the peculiar range of the botanist; and even now we may hear them termed sea-weeds by persons ignorant of their real nature," though "they are creatures endowed with vitality, and with various powers peculiar to [animal] life."[36] "To look on such objects with indifference, is indeed willfully to disregard the beautiful and the wonderful among the works of God," the writer explains.[37] Coral matters to this Methodist author, then, because it warns us against what Emerson might call an "impudent knowingness" that is actually ignorance—whether of coral or any other "works of God." Readers could even practice *not* mistaking coral for a plant by gazing at the incredibly tree-like coral branch embossed on the text's cover (fig. 3.1) and avoiding the errors that come from evaluating nature by its form.

In a final example of evangelical interpretations of coral, American author and Congregationalist minister John Lauris Blake (1788–1857) states more directly that coral encourages us to reconsider our ranking of other groups of beings. In *Wonders of the Ocean* (1845), Blake includes the following reflection, which nearly anticipates Schele de Vere's analogy of corals to "other organic forms" in his later and similarly titled *Wonders of the Deep*. Blake writes,

> An animal barely possessing life, scarcely appearing to possess volition, tied down to its narrow cell, ephemeral in existence, is daily, hourly, creating the habitations of men, of animals, of plants. It is founding a new continent; it is constructing a new world. . . . Yet man, vain man, pretends to look down on the myriads of beings equally insignificant in appearance, because he has not yet discovered the great offices which they hold, the duties which they fulfil, in the great order of nature.[38]

Like the author of *The Coral-Maker*, Blake does not specify which "beings" are presumptuously undervalued by "vain man" because they are "insignificant in appearance." Yet like Margaret Coxe, Blake also supported the social elevation of women in particular. In 1828 he had founded *Ladies' Magazine*, the nation's first magazine edited by and for women. The first woman that Blake hired to edit the magazine was Sarah Josepha Hale.

Today Hale is popularly remembered for editing *Godey's Lady's Book* and authoring the song "Mary's Lamb" (1830), otherwise known as "Mary Had a Little Lamb." But Hale also wrote a song about a coral specimen, "The Coral Branch" (1834), which was embraced by thousands of Americans, from a variety of social classes, for several generations, as its robust publication and performance history attests.[39] Hale's song preserves and transmits Peyssonnel's discovery of the coral insect. It frames everyday encounters with coral as a chance to refuse the essentializing logic that had kept generations ignorant of coral's true nature and that continued in Hale's time to rationalize the extractive

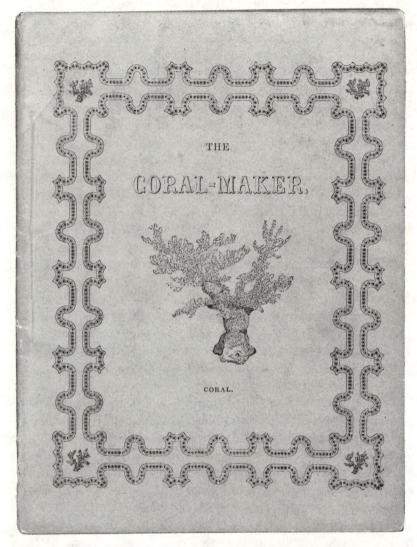

FIG. 3.1. Evangelical coral: This small, square pamphlet, a natural history of the sea and evangelical tract issued by the Sunday School Union of the Methodist Episcopal Church, features a coral branch on the cover. The text explains, "To look on such objects with indifference, is indeed willfully to disregard the beautiful and the wonderful among the works of God." Cover of *The Coral-Maker*, 1842 (Bodleian Library).

labor of some groups. And it draws from reefs the lesson that a human collective thrives when people choose work that sustains bonds, past and present. Here are the lyrics of Hale's song, as it appears in her *School Song-Book* and as scored by Boston musician George James Webb (fig. 3.2).

> I thought my branch of coral
> A pretty shrub might be,
> Until I learned a little worm
> Had made it in the sea—
> Down, down so deep,
> Where dark waters sleep,
> The coral insect lives;
> But rests not there,
> With toil and care,
> It upward, upward strives.
>
> It builds its coral palaces,
> Than lofty hills more high;
> And then, the structure to complete,
> The little worm must die,—
> Thus teaching me,
> When coral I see,
> That, dying, I should leave
> Some good work here,
> My friends to cheer,
> When o'er my tomb they grieve.[40]

Hale's song opens with the (by then) familiar rejoinder that coral is not a member of the vegetable kingdom, although it looks like one. In fact, coral appears so plantlike that the speaker "thought" her "coral branch"—presumably a coral specimen of the kind that popularly decorated mantels and curiosity cabinets—was a "pretty shrub," until she learned that a "coral insect" made it "down, down so deep" in the sea. As generations of Americans recited these lyrics, they were not only acknowledging coral's animal nature, in spite of appearances; they were doing so in the very language that Peyssonnel uses in *Traité du Corail* where he announces that coral is the production of "the coral-insect." The speaker follows that recognition with the realization that coral is in fact both the "structure" produced by *and* the bodily remains left behind by "the coral insect," but that realization is not cause to imagine that coral insects labor repetitively and inexhaustibly to produce a structure that consumes and

### THE CORAL BRANCH.

A JUVENILE SONG BY MRS. S. J. HALE.

MUSIC BY G. J. WEBB—WITH AN ACCOMPANIMENT FOR THE PIANO FORTE.

FIG. 3.2. Singing about coral: Boston composer George James Webb arranged a musical score for Sarah Josepha Hale's "The Coral Branch," which was sung or recited by thousands of Americans across several generations in schoolrooms, at home, and in public church concerts. George James Webb, "The Coral Branch," 1835. *American Annals of Education and Instruction: Being a Continuation of the American Journal of Education.* To hear a recent interpretation and recording of the song, see "The Coral Branch." Michele Navakas (personal website): https://michelenavakas.com/thecoralbranch.

forgets them (as Sigourney and other writers imagine in the reflections on coral considered in the previous chapter).[41]

The fact that coral insects "upward, upward strive" to build "coral palaces" from their bodies is cause to reimagine the enduring value of the speaker's own life and the particular form of labor she pursues: "When coral I see," she reflects, it teaches me to pause, remember reefs emerging from polyps, and then "work" in ways that sustain a grieving community of "friends" long after my death.[42] Put another way, the lesson of coral is that people, too, should take up labors that bind past and present—that commemorate particular lives now past, while sustaining a living community.[43]

It is easy to envision Hale's speaker, and those who recited her first-person lyrics, passing by a coral branch repeatedly in the course of any given day. Branches of coral were commonly collected and displayed on mantelpieces or in curiosity cabinets in nineteenth-century US homes. Such specimens were also popular subjects of stereo cards that allowed people to peer deeply into coral, such as the one owned by a friend of Hale's, Oliver Wendell Holmes (the two met while living in the same Boston boardinghouse; fig. 1.22).[44] Hale's song frames these familiar specimens as common occasions to practice peering closely, while remembering that looking could not always discern, pin down, or predict the identity and value of what one sees.

Hale wrote frequently about coral in several genres. Coral is the chief subject of a chapter titled "The General Resemblance of Vegetable and Animal Life" in an extremely popular natural history that Hale abridged and edited for a US audience, British author John Mason Good's *Good's Book of Nature* (1834).[45] Here Hale frames coral as the most obvious instance of nonhuman nature that looks like a taxonomically lower form of life yet turns out to be a higher one. "At the first glance, it seems very easy to tell the difference between a vegetable and an animal," the chapter begins.[46] "Yet there are things formed, which it is very difficult to name, and we hardly know, whether to call them animal, vegetable, or mineral. Coral is an animal substance; but it looks like a vegetable and feels like a mineral."[47] That idea is reinforced by reading comprehension questions that prompt readers to ponder the determinative value of visible features: "What is the difference between a vegetable and an animal? Between a vegetable and a mineral? Is it always easy to distinguish these? . . . Can you name any other substance which it is difficult to distinguish?"[48] Coral's history raises questions that radiate outward from coral, encouraging readers to wonder what else they had failed "to name" and to "know" accurately.

Central to the very definition of coral in the nineteenth-century US, then, was its storied defiance of a system of taxonomic ranking adjudicated by reference to the visible. If one could go back in time to ask even a modestly educated nineteenth-century American "What is coral?" they would likely answer correctly that coral is an animal. Yet many might also eagerly recite the doubt, dispute, and denial of coral's animal status that had erupted during earlier centuries. Some might even recount the specific names and claims of European naturalists who had failed to perceive coral's nature and capacities based on its appearance. That particular failure of perception was encoded in the familiar coral specimens, coral jewelry, and writing about coral that regularly circulated through American hands and homes, so that coral frequently marked the limits of essentialist determinism just as that ideology was becoming, as sociologist Collette Guillaumin observes, "the chief victory of the new scientific spirit."[49] Amplifying the power of coral to defy human perception and prediction during the mid-nineteenth century was another and more recently emergent domain of knowledge about coral.

## FOUNDED IN RELATIONS: DARWIN'S "CORAL INSECT" (1842)

When Darwin first encountered a coral reef in Tahiti in 1835, he was fascinated by the "almost gelatinous bodies of [the] apparently insignificant creatures" that combine their "vital energies" with one another, the sea, and multiple "other and unexpected agents at work"—including "detritus," "sandstone," "shells," "the spines of echini, and other organic bodies"—to produce, sustain, and expand "the solid reef."[50] What mattered to most scientists about Darwin's theory of coral formation was its geological implications, for Darwin showed that the location and shape of reefs indicate whether and where the earth's crust is either sinking or rising, a process that usually happens so slowly that humans cannot perceive it.[51]

What mattered to the US writers and activists whose work I will consider in the next chapter was mainly that reefs, on a Darwinian account, emerge from so many relations that humans cannot predict the future form of any given reef in process and that the tiniest movements of polyps, though apparently inconsequential, could alter the entire structure over time. While Darwin never called coral polyps "coral insects," instead referring to them in the treatise as "polypifers" or "the corals," Americans frequently identified Darwin's reef-building polyps as "coral insects," according to discussions and reviews of the coral treatise in popular US publications.[52] In Darwin's account of reef formation, Americans found another account of coral insect life and labor during a period when people were already taking political lessons from coral insects on how to form the most robust human polity.

According to Darwin's treatise, reefs are established, sustained, and expanded by multiple spatial and temporal relations enabled by polyps. Throughout *The Structure and Distribution of Coral Reefs* Darwin describes polyps as continually "building" and "constructing the reef" by way of interactions with innumerable other entities and influences, living and nonliving, so that "the relations . . . which determine the formation of reefs on any shore, by the vigorous growth of the efficient kinds of coral, must be very complex, and with our imperfect knowledge quite inexplicable."[53] Or as Darwin also explains, "We can by no means follow out all the results" of the ongoing process of reef formation.[54] It's all constant "round[s] of decay and renovation," and while a coral reef or island may expand, it could also stay the same size, or wear away.[55] Reefs, then, are always becoming something other than what their past and present forms could allow us to predict because they are determined by polyp relations, across time, with other polyps, variable currents and winds, sand and ocean detritus, and other species and their remains.[56]

Viewed in this way, Darwin's coral treatise anticipates the conception of all life that he would later elaborate more fully in *Origin of Species* (1859) and other writings. That conception, philosopher and feminist theorist Elizabeth Grosz explains, is characterized by "a fundamental *indetermination*," subject to wide-ranging influences generated by individual variations, "open-ended," "unpredictable and inexplicable in causal terms," "emergent," and always potentially "otherwise than its present and past forms."[57] Such a conception "offers a subtle and complex critique of both essentialism and teleology," Grosz demonstrates, for on this account of life, biology continually intermingles with and responds to material and historical forces, while temporal movement forward does not automatically constitute development or progress.[58]

What I am suggesting is that nearly two decades before *Origin*, Darwin introduced what Grosz calls a "conception of life as ceaseless becoming," as "dynamic, collective change," through coral.[59] In the US Darwin's reflections on coral were widely publicized beyond scientific circles, since popular lectures, sermons, and periodicals—from *Godey's Lady's Book* to the *Christian Register*—disseminated Darwin's account of reefs to an interested public.[60] As Darwin's treatise made its way into US culture, it produced a new conception of coral reefs and islands, one that differed markedly from that which inspired the writers considered in the previous chapter to describe reefs as systems that thrive most by consuming and excluding one group of beings for the sake of another (while sometimes appearing to sustain all parties through individual labor for individual gain).

Before Darwin, it was easier for Americans to imagine coral islands as material evidence that the pursuit of self-interested labor was politically progressive.

This is largely because one of the most popular descriptions of coral islands in nineteenth-century US print culture attributed a distinct teleology to their particular mode of growth. That description was penned by British explorer Matthew Flinders in the second volume of *A Voyage to Terra Australis* (London 1814), Flinders's narrative of his 1801–3 circumnavigation of Australia, a text that Darwin also cites in his coral treatise.[61] Flinders's coral island inspired poet James Montgomery's description of Pelican Island in his 1826 epic—beloved by US readers, and considered in chapter 2—and it appeared in one form or another across countless US reference works and periodical essays that defined coral islands by citing, quoting from, or in some way drawing on Flinders's description, though frequently without attribution.[62] Flinders's coral island differs vastly from Darwin's account of nonteleological change by way of "vigorous" polyp relations.

According to Flinders, coral islands begin in the labor and death of polyps (which he calls "animalcules") and then progress inevitably upward and outward until they achieve the end goal of becoming humankind's property. A coral island starts to emerge "when the animalcules which form the corals at the bottom of the ocean, cease to live," at which point their remains "adhere to each other" in an ever-expanding mass whose "interstices" are "gradually filled up with sand and broken pieces of coral washed by the sea."[63] And then "future races of these animalcules erect their habitations upon the rising bank, and die in their turn to increase, but principally to elevate, this monument of their wonderful labours." Once "their wall of coral" rises to break the sea's surface, the corals "[send] forth" "infant colonies" to die and attach in turn, as multiple other forces and agents begin to contribute to, and benefit from, the island's growth above the sea.[64] And then comes the part of Flinders's description that was most enthusiastically embraced by US readers:

> The new bank is not long in being visited by sea birds; salt plants take root upon it, and a soil begins to be formed; a cocoa nut, or the drupe of a pandanus [the fruit of a tropical pine] is thrown on shore; land birds visit it and deposit the seeds of shrubs and trees; every high tide, and still more every gale, adds something to the bank; the form of an island is gradually assumed; and last of all comes man to take possession.[65]

Taking up Flinders's description in *Pelican Island*, Montgomery adds more to the heap—wrecks, exuviae, shells, pregnant rocks, sloughs, bones of sea-monsters and whales, weeds, and "unutterable relics"—yet in both texts, and in the array of US sources that Flinders also inspired, the "various mass" (as Montgomery calls it) somehow always grows predictably and progressively

toward a form of human flourishing grounded in territorial possession.[66] As
a small sampling of US writers influenced by Flinders expressed it: "Last of
all comes Man, and the island forms a part of the inhabited world."[67] "After a
succession of ages, man would take possession, and a populous country in time
be formed."[68] "Thus the soil comes to be inhabited; and man at last comes,
and forms a settlement."[69] It is not difficult to see how the popular version
of coral islands installed by Flinders lent itself to the promotion of a politics
of inevitable progress grounded in individual labor for acquisition and self-
sustenance (while actually relying on the wholesale extraction of one group
for the sustenance of another).

Yet Darwin cannot say with Flinders and Montgomery that a coral reef
is destined to become man's ever-expanding landed possession. It would be
impossible to bolster the claims of capitalist individualism with reference to
Darwin's coral island because we cannot know what precise form any given
coral island will assume through polyp relations with entities and forces across
time, or even whether the island will vanish into the sea. Darwin's reefs thereby
pose a challenge to teleological growth, and to an exclusively biological account of life, challenges that Grosz locates in Darwin's later writings. In reference to these writings, Grosz wonders why feminists have not more fully
analyzed and embraced "the social and political implications" of Darwin's
conception of life.[70]

In the next chapter I will suggest that they did during the nineteenth
century, not in the wake of *Origin* but rather in the wake of Darwin's coral
treatise. As an emergent Darwinian conception of reefs circulated alongside
the taxonomic crisis raised by coral's natural history, a new frame of reference
for coral and coral islands, coral insects, and coral objects emerged. In the
unpredictable, indeterminate growth by relations enabled by seemingly insignificant entities, many Americans found grounds for hope—and instructions
for building—a different future.

INTERCHAPTER
# "I Come from Coral Reefs"

One origin of the following chapters is a single line in Guyanese poet Grace Nichols's "Web of Kin," which appears in her acclaimed first book of poetry, *I Is a Long Memoried Woman* (1983).[1] While reflecting on experiences of slavery particular to Black women, the poem's female speaker, a survivor of the Middle Passage, declares "I come from coral reefs."[2] By proclaiming her coral origins and nature, she announces her "oceanic experiences" along with her "hybrid and transitory form of existence," alive and dead, African and Diasporic Caribbean, and continually growing and changing in concert with other organic forms, as scholars of the poem have demonstrated.[3]

In this way Nichols participates in the complex relationship to ocean ecology that literary scholar and poet Joshua Bennett identifies among other Black writers throughout the African Diaspora: for Nichols's speaker acknowledges the formative role of transatlantic slavery and Middle Passage, yet refuses to allow that history to fully determine her relationship to "the sea and its animal lifeworlds," instead finding grounds for identification and imagination with "the lives of the nonhuman animal entities that dwell within the oceanic realm."[4] By claiming that she comes from coral in particular, Nichols's speaker may be claiming not only a transitory and hybrid identity, however, but one that Bennett might also consider "unfathomable, untamable, endless," since these specific qualities have long been imagined as defining features of coral.[5]

The speaker's complex relationship to coral in particular set me on the path of looking for nineteenth-century US versions of that relationship—that is, instances in which marginalized persons claim kinship or identification with coral during a time when coral's various colonial histories and meanings were all the more evident, and not only because of its uniquely oceanic nature and identity. Coral had also recently served as currency in the transatlantic slave trade, and as a specimen relentlessly scrutinized and taxonomized by white European natural historians. Moreover, it had become a popular metaphor that normalized different forms of coercive labor and elided Black lives in particular (see chapter 2). What would it mean for a nineteenth-century African American writer or activist to identify with coral or even to claim that their body or community is like coral? For that matter, what would it mean for a nineteenth-century US white woman to do so, albeit from an entirely different marginalized relation to the polity?

Using Nichols's speaker as a guide to the prospect that coral's colonial histories could not eclipse its anticolonial possibilities, even during the nineteenth century, I studied the archive more closely. I looked for moments when differently marginalized persons took up coral to challenge the totalizing imperatives of slavery and other colonial projects. I looked for evidence that they drew on coral to imagine and announce identities (personal or communal) that exceeded colonial frameworks of meaning. I tried to find instances where they harnessed, to their own ends, a Darwinian sense of reefs as structures forged through relations and thus always becoming otherwise than their past and present forms. The following chapters are the results of that search.

# 4
# Coral Collectives

We, the corals of another era, growing out of each other,
elaborating new conditions of existence. . . .

—Elizabeth Wormeley Latimer, "Coral Creations" (1856)

In "Coral Creations" (1856), historian and translator Elizabeth Wormeley Latimer makes a specific political claim about the US by evoking a Darwinian
account of reef formation as an indeterminate, relational process. The country,
Latimer claims, is sustained by people she imagines as "the corals," tiny bodies
that labor to produce and support a larger collective by forging interpersonal,
intergenerational bonds.[1] "We, the corals," Latimer calls these bond-builders,
echoing the Constitution's "We the People," while being much more particular
about exactly who is this "we."

For Latimer, corals are those persons who presently perform the most
constitutive labor, silently and at personal cost, to build a country to which
they might never fully belong. They are also the ones whose work might just
transform the US into a more inclusive structure. For through their social,
economic, and political binding of persons, corals are also ever "elaborating
*new* conditions of existence"—the word "elaborating" meaning both to design
and to make.[2] Enormous changes to the entire system, in other words, might
emerge from relations, past and present, enabled by the tiniest movements of
individual bodies so small as to be invisible; Latimer suggests that the human
equivalents of those bodies are women.

By drawing on Darwinian principles of reef formation to recognize the
work of women—and to call on others to join women in that work—Latimer
participates in the overlooked literary and cultural phenomenon that this chapter explores: during the nineteenth century differently marginalized persons,
white and Black, imagined and practiced "coral collectives," my term for their

coral-inspired alternatives to a long-standing version of political common good grounded in the pursuit of self-interest.[3]

The chapter begins by examining a popular nineteenth-century political use of coral that rests on a pre-Darwinian, teleological understanding of coral islands and that all the writers, speakers, and activists assembled in the rest of the chapter challenged: the claim that, by merely living together under US democracy, the country entire had now become a coral island that grows progressively by way of mutual sustenance, all individuals continually enriching all others in equal measure by each accumulating their own material gain. The writers who subscribe to that claim seem to believe that merely *describing* the US as a coral island makes it so.

The chapter then turns to those who reject that claim and instead *exhort* people to act in ways that could eventually render the US as mutually sustaining as a Darwinian coral island, a process that must begin by recognizing that presently only some of us are the bond-building polyps. For these writers, reefs do not reflect the present conditions of mutual sustenance and liberty already achieved in the US or guarantee that future political progress depends on self-interested labor. Rather, reefs model an ideal collective toward which all of us must continually strive by taking up labor aimed at recognizing and redressing present inequities and building a less exclusionary polity.

Many white women's writings on coral, labor, and women echo Latimer's call for "new conditions of existence," including work by feminist and philosopher Ellen M. Mitchell, by Boston adoption reformers Anstrice and Eunice C. Fellows, and by abolitionist Harriet Beecher Stowe. These women told themselves and other women that the labor women carried out in parlors, nurseries, and other less visible spaces was indispensable to the country's social and economic thriving, and they demanded that more people take up that work. In this way, they declared the indispensable role of various types of women's work to the emergence of industrialized society in the US. While historians have verified that indispensability, literary scholars struggle to find evidence that women themselves both grasped and expressed the crucial role that they played, largely because the pressures to erase the value of women's work could not have been more intense.[4] As literary scholar Elizabeth Dillon (2004) has shown, the ideal of the autonomous liberal male, buttressed by the foundational myth of American exceptionalism—by which the individual pursuit of economic self-interest sustains the whole of society—requires the continual erasure of women's bodies and labor. It is not surprising then that women participated in representing their labor as marginal; it was not uncommon for them to "pastoralize" housework, for example, refashioning arduous, daily labor as merely "a mode of feminine being or affect."[5] By way of coral,

PLATE 1. Bedecked in coral: This Baltimore child wears a necklace of red coral beads, matching coral bar pins at each shoulder, and a silver "coral and bells" on a long gold chain around her neck. Joshua Johnson, *Emma Van Name*, 1805 (The Metropolitan Museum of Art).

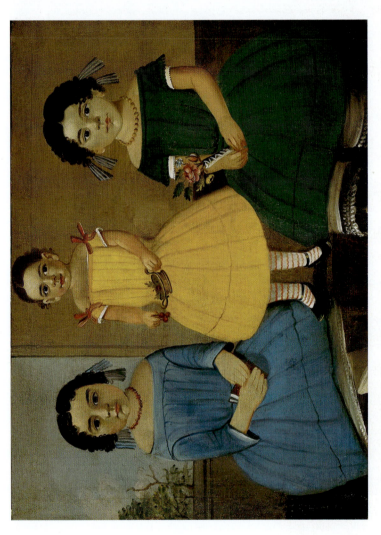

PLATE 2. A red coral necklace: In this portrait of three sisters, daughters of African American real estate investor Samuel Copeland of Massachusetts, the child at left wears coral. William Matthew Prior, *Three Sisters of the Copeland Family*, 1854 (Museum of Fine Arts, Boston).

PLATE 3. Allegorical coral: While free persons of African descent in colonial Dutch Brazil might have actually worn red coral necklaces, this portrait may be more allegorical than documentary, considering the European artistic tradition of portraying the continent of "Africa" as a woman wearing or holding red coral. Albert Eckhout, *African Woman and Child*, 1641 (National Museum of Denmark).

PLATE 4. Teething on coral: A combined teething aid, rattle, whistle, and talisman, the "coral and bells" was a popular gift for babies of all genders in more affluent, usually white families. This one belonged to US poet Henry Wadsworth Longfellow, who gave it to his niece in 1852. "Coral and Bells," ca. 1850 (Courtesy of National Park Service, Longfellow House-Washington's Headquarters National Historic Site, LONG 27636).

PLATE 5. Coral as fertility object: In this portrait of the Winslow family of Boston, Isaac Winslow's wife holds their younger daughter Hannah while grasping a coral and bells. Notice how the object echoes the shape of Winslow's finger and points downward to the shadowy region beneath the uplifted skirt of their elder daughter. Joseph Blackburn, *Isaac Winslow and His Family*, 1755 (Museum of Fine Arts, Boston).

PLATE 6. Reef specimen: This drawing represents a red coral reef specimen that zoologist James Dwight Dana brought back to the US, likely from Fiji, when participating in the US Exploring Expedition (1838–42). The specimen itself was publicly displayed in Washington, DC, at the Patent Office Museum. James Dwight Dana, *Zoophyte Atlas*, 1849. Plate 60, figure 1. "*Corallium secundum*, natural size."

PLATE 7. Making a living from coral: On a trip to the Bahamas in the mid-1880s, US painter Winslow Homer portrayed local Black divers retrieving coral specimens, which they could then sell to tourists. This diver seems to show off his find, possibly as a gift to the woman aboard the boat. Winslow Homer, *Sea Garden, Bahamas,* 1885 (Harvard Art Museums).

PLATE 8. Bringing coral from below, this diver seems ready to hand it up to two fellow workers waiting aboard the boat. Winslow Homer, *The Coral Divers,* 1885 (Private Collection).

PLATE 9. Creole identity: A red coral necklace is worn by this unidentified New Orleans Creole and free woman of color. Red coral ornaments may have expressed status and Creole identity among Creole women of African descent in nineteenth-century New Orleans. Louis Nicholas Adolphe Rinck, *Woman in Tignon*, 1844 (Hilliard Art Museum).

PLATE 10. Topsy wears a full set: Notice the coral jewelry at Topsy's ears, neck, and wrist in this illustration of a scene from Harriet Beecher Stowe's *Uncle Tom's Cabin*. Illustrators of the novel may well have been influenced by the allegory of "Africa" as a woman wearing coral: compare this portrait to Eckhout's *African Woman* (shown in Plate 3). J. A. Bingham, "Topsy and Eva," 1857 (Harriet Beecher Stowe Center, Hartford, Connecticut).

however, many white women communicated their value and imagined their labors as the essential basis of a more inclusive collective—albeit one that did not necessarily include everyone.[6]

As I show in the third and final part of this chapter, the most expansive reef visions of even the most radical white women writers still foundered on race, which is partly why Black writers and activists repeatedly drew on some of the same conceptions of coral to articulate their own visions of coral collectives, which could expand beyond Black communities only if white persons actively recognized and redressed racial inequity.[7] Frances Ellen Watkins Harper, James McCune Smith, Jonas Holland Townsend, Ferdinand Lee Barnett, and Abraham Lincoln DeMond evoked coral to call on fellow Black Americans to take up labors aimed at building communal bonds, past and present, as the necessary foundation of hope for a more vibrant future.[8]

It is crucial to recognize that Black writers who claimed coral as a political metaphor for recognition and collective building both dared and risked much more than white women who did the same. For one, these Black writers worked from and toward very different conditions: slavery and its legacies are in no way comparable to the unpaid domestic and maternal roles traditionally performed by white women.[9] And these Black writers also faced another challenge: the coral metaphor had long been used by white writers to reduce Black persons to laboring bodies, conceal the extractive labor conditions of slavery, or elide Black persons from history altogether, as I show in chapter 2. By evoking coral to affirm their presence and contributions, then, these Black writers faced a challenge that Lindon Barrett describes with reference to Black writers who used the conventions of autobiographical writing to narrate their experiences of slavery: they dared to talk about their spatial, material, and spiritual presence and importance in a medium—in this case, the popular discourse of coral, labor, and politics—that had historically been "necessarily antithetical to that project."[10] One sign that they succeeded is the uptake of coral by an actual Black collective in the 1890s: the interchapter tells the untold story of the Coral Builders Society (ca. 1892–1901), an auxiliary of Cleveland's St. John's AME Church that named themselves after coral and called for more people to reform the reef.

All of the people—Black and white—centered in most of this chapter responded to various political and social inequities by drawing on the nature and histories of coral to illuminate the unjust constraints and exclusions that defined their present political and social positions; to tell one another how to work *from within* those positions and *toward* a world less divided by colonialism, class, and gender; to reject self-interest as the foundation of collective welfare; and to call for wider acknowledgement of, and participation in, a slow, transformative work of binding that they considered the necessary foundation of a social justice that had yet to materialize within the US.[11] They theorized that a more just society *might* emerge if more people assented to, and participated in, an alternate version of collective welfare that could only come from the kinds of bond-building labors that had thus far fallen almost entirely on differently marginalized groups to perform. Collectively their work affirms that one crucial political value of coral has been its power to expose the republic's extraordinary demands on laboring bodies, even—and perhaps especially— when we try to remain silent about those demands. As we try to imagine more just systems of human labor now, a past infused with coral crucially reminds us how easily a system of labor appearing to offer social and economic justice can silently extract the life and labor of the many for the benefit of the few. It also offers us better prospects that we can still put into practice.

## "Society is that Island"

A brief, anonymous essay titled "A Branch of Coral" (1860), which appears in the Quaker magazine *Friends' Intelligencer*, describes a reef as the ideal polity:

> Here is a branch of coral, which you know to be in its living state a colony of polypes. Each of these multitudinous polypes is an individual, and each exactly resembles the other. But the whole colony has one nutritive fluid in common. They are all actively engaged in securing food, and the labors of each enrich all. It is animal socialism of the purest kind—there are no rich and no poor, neither are there any idlers.[12]

According to this writer, a reef is a glorious manifestation of mutual self-sustenance, a structure in which each individual sustains all others merely by sustaining oneself, for "all [individuals] actively" labor, and "the labors of each enrich all" as "one nutritive fluid in common." There are no economic disparities—"no rich and no poor"—and indeed no disparities at all, for in this society "each [individual] exactly resembles the other."

According to a number of nineteenth-century writers who analogized the US to a reef, the country had now become the very coral island described in "A Branch of Coral." So imagines the Reverend Henry Ward Beecher, a prominent Congregationalist minister and the brother of Harriet Beecher Stowe, when speaking from the pulpit of Brooklyn's Plymouth Church in 1868: "Man is like the coral builder, which is a little worm, in its own little cell, doing its own little work, adding its own little substance to the work of others, and dying where it began, leaving the reef somewhat enlarged by what was only a selfish architecture."[13] Beecher is not exactly advocating that "man" engage *only* in his "own little work"; yet that "selfish architecture" does "enlarge" the reef. Beecher's analogy of "man" to the "coral builder" suggests that society, like a reef, is inevitably sustained, and even productively expanded, by each person's pursuit of self-interest.

Exactly how that happens is the subject of an earlier sermon (1863) in which Beecher evokes a coral island to support the claim that the untrammeled pursuit of individual wealth and private property is a great social good:

> I have seen men carrying waves of wealth, and before they had gone far one wave would swallow up and devour another. Wealth is unstable. It is changing hands perpetually. Men are earning and losing. Only society never fails. Society gains, whoever goes down of its individuals. . . . The island that the coral insect makes grows. The insect is ousted, time after time, from his little tenement; but he leaves behind him what he has built to swell the bulk and stability of the island. And society is that island, which men are building up by their industries. It does not become bankrupt, nor fail, but the men that build on it are constantly fluctuating and changing.[14]

Beecher's point is that, just as whatever "the coral insect" builds for itself inevitably "swell[s] the bulk and stability" of the island as a whole, so whatever "waves of wealth" individual humans generate inevitably build up "society." Beecher's wealth "waves" work much like the "one nutritive fluid in common" in "A Branch of Coral," flowing into each person and thereby (somehow) into all other persons evenly. And crucially, Beecher's island always progresses, just like the coral islands described by Flinders and Montgomery that I examined in the previous chapter. As long as each of us pursues some kind of material accumulation, society "does not become bankrupt, nor fail" but rather "gains, whoever goes down of its individuals."[15]

A coral island supports similar political claims in "A Mutual Dependence," an essay by US novelist Harriet Prescott Spofford that appeared in a self-help guide called *Stepping-Stones to Happiness* (1897). The essay argues that "no one lives in the civilized world who is not doing something for some one else, either physically or intellectually or spiritually."[16] As an "illustration" of how "all humanity works together" by each "member" working for itself, Spofford offers the example of "the wealthy woman stepping from her stone mansion to her carriage." "To how many workers has she contributed a fractional support?," Spofford asks before listing a dizzying array of persons whom this one woman is "lifting up" by merely existing.[17] "For her" labors the quarryman, kiln worker, woodman, miner, mason, blacksmith, marble worker, plumber, watchman, street cleaner, lamplighter, doctors, slaughterers, leather dressers, carriage makers, coachmen, footmen, Texan vaqueros, South American fishermen, "Hindoo" farmers, European peasants, "the negro bent under the sun picking the cotton," "Oriental" weavers, diamond miners, sailors, traders, railroad workers, journalists, firemen, engineers, military troops on "the frontier haunted by the tomahawk," musicians, dancers, actors, judges, and washerwomen.[18] "She cannot do without them, as they can not do without her," reflects Spofford, for "her life is their life," and however wealthy she is, to them she is a "bound woman."[19] To shore up her argument, Spofford concludes with the analogy of a coral reef, which materially illustrates our "intimate" and mutually sustaining bonds: "Let us be ever so much accountable to fate and to our consciences as separate individuals, we are yet more certainly congregated and bound together in one great circulation and interchange than the atoms of some vast polyp building its coral reef in the South Pacific, and every one's self-respect and reverence must have its effect upon the individuality of every other soul."[20]

The reef at the end of Spofford's essay resembles Beecher's coral island: both writers ground their claims in a distinctly pre-Darwinian conception of coral, which enables them to insist that society's progress is the inevitable outcome of each person just doing their thing, almost whatever it is. Yet "A Mutual Dependence" allows us to see that this vision of a thriving democracy of freedom and reciprocity depends on the suspension or erasure of race, class, gender, and other differences, as well as the political and social inequities that historically attend those differences. For when Spofford says that people are as "congregated and bound together" as the atoms of a coral polyp, she means something like "we're all in this together," which is to say that the constraints binding the poor Black cotton picker "bent under the sun" to the person who owns the land are the same as those that "bind" the "wealthy woman" to the washerwoman. Like the very atoms of a single polyp, we are all mutually enriched by "one great circulation and interchange," which might flow like Beecher's "waves of wealth" or the common "nutritive fluid" of "A Branch of Coral." Spofford seems to be using the case of coral to announce that the US is a place where everyone accepts, or just stops worrying about, even the most radical disparities—economic and otherwise—which are anyway the essential condition of mutual thriving, including liberty, political stability, and social progress.[21]

Collectively these sources attest to the ease with which a pre-Darwinian coral reef analogy could render the country's profound inequities both natural and harmless. They suggest that visions of the US as a coral collective—a polity in which social stability and progress flow naturally from our already existing interdependence—could transpire by ignoring or embracing the historical and ongoing disparities that, in reality, profoundly constrain and silence some persons so that others can move more freely. As Mary Tyler Peabody Mann, wife of education reformer Horace Mann, would explain in an essay called "Coral" (1880), a reef consists of "separate individuals"—including polyps, anemones, jellyfish, seaweed, water, and lime—"growing . . . all upward together."[22]

The political appeal of this "all upward together" is powerful; yet in historical reality we know that society is not that island and to insist that it is leaves no room to lodge a complaint. What, one must ask, should a person do who perceives that we are not all mutually "congregated and bound" together, persons of all races, classes, and genders, each continually swelling our society island toward greater progress and stability as we pursue our own "selfish architecture"? The coral collectives of Beecher and Spofford cannot entertain such a question, let alone furnish an answer.

But is that growth lively and active, or just a slow, steady accretion of dead matter? In yet another sermon (1871), Beecher describes reefs as remains. Discussing the deadness of "torpid" ideas, he explains that "they are almost like the coral; they have received a fixed form. They have lost their life, but the form remains."[23] Beecher's language captures the stasis and suspension tacitly recommended, and even ensured, by the political understanding of coral that we have been examining. These visions of human polities as coral collectives promise, and even appear to resemble, lively societies actively working together toward mutual growth—"all actively engaged," to quote "A Branch of Coral"— while in fact they tell individuals within that society to be silent and accept their ostensibly preordained, fixed, and often highly constrained positions by renaming those positions freedom. Put otherwise, these reef-inspired visions of human society leave no directions for how to act toward attaining equality and interdependence, for they presume that such ideals are built into the very nature of US life and are thus ever-increasing for all in equal measure.[24] Yet Americans also had other sources of knowledge about how coral grows.

The women whose work features in the next section of this chapter endorsed the claim that we have been exploring, namely that coral islands grow by mutual sustenance, all individuals continually enriching all others. Yet these women also contested the assertion that such growth had already been achieved among humans and would only continue to flourish, merely by living together and pursuing our "selfish architecture." To these women, the political value of coral islands was not that they reflected back to society the current or future conditions of the growing polity; rather, coral islands embodied an ideal toward which all humans, differently positioned, must continually strive through a particular form of labor, while never presuming to have achieved that ideal.

## We Are Not That Island Yet

In a coral island Elizabeth Wormeley Latimer saw what Beecher and Spofford also saw: a natural example of a society in which each individual sustains all others, the good of the whole dependent upon the labors of each. Her essay, "Coral Creations" (1856), tells us what we need to do if we want to build that coral country. For presently, according to Latimer, only some of us are coral polyps. After a definition of reefs as collective organisms constituted over time by tiny laborers, Latimer concludes her essay with the following political reflection:

> The coral animals are so small as to be almost invisible, yet are the architects of continents. If we could penetrate the mysteries of human history, we should find . . . a world changed for all time by the movements of persons who cannot indeed be forgotten, because they were never known . . . We, the corals of another era, growing out of each other, elaborating new conditions of existence, have an end to fulfil in the doings of Providence, and no substitute can be found capable of occupying our place. Though humble amid the wonders of the universe, we are by no means unnecessary to the accomplishment of its conditions . . . we should . . . remember that we too exist for a purpose, and assume the relations devolving upon us with modesty and discretion, yet with energy and boldness.[25]

A new, complex claim sets Latimer's reef politics apart from those of the writers examined in the previous section of this chapter and aligns her with many other nineteenth-century US women whose work features below: if we want to transform the US into a coral collective—a more equal and mutually sustaining polity—then more must do the difficult, less visible labor ("elaborating") that presently only some are doing, namely, the labor of binding persons together across certain differences and within and among generations.

First and foremost, Latimer's phrase, "We, the corals," echoes the Constitution's "We the people," while altering its meaning in a number of significant ways. "We, the corals" does not ask us to buy into an abstraction that ignores the real differences (social, political, economic) that presently constrain and/ or silence some people while privileging and rewarding others. For Latimer acknowledges that the constitutive "we" presently consists only of certain groups of people. Corals are those whose labor goes unrecognized, "never known" in the pages of the very "human history" they propel. They are the bond builders, "building" "relations" across and between generations, each individual simultaneously connecting itself to its contemporaries, deeply "rooted" in, and "growing out of," the bodies of the previous generation and producing and sustaining the next as they "shoot out new offspring."[26] Those bonds are not exclusively biological. Coral progeny are born into and sustain a community of other corals and additional "contributors," whom Latimer mentions earlier in the essay, including shellfish, sand, and driftwood. And though "almost invisible," "small," and "humble," corals are indispensable to the polity entire, for they "are the architects of continents."

"Coral Creations" appeared in Boston's *Waverley Magazine*, a periodical that attracted authors "denied space within other [periodical] pages" due to race, class, and/or gender.[27] It seems likely then that Latimer's readers would have identified her "we" as socially and politically marginalized persons, those "persons who cannot indeed be forgotten, because they were never known." If that is the case, then Latimer is exhorting these persons to labor from within their presently constrained conditions in ways that might create what she calls "new conditions of existence." For her final sentence is an exhortation to nourish interpersonal bonds, which is what Latimer means when she writes that "we should . . . assume the relations devolving upon us with modesty and discretion yet with energy and boldness"—the word *assume* here meaning to accept responsibility for, and take on, and possibly even take into, the body. That work is the difficult and uncertain work of "elaborating" (to design and to make) through which the world might be "changed for all time" by "the movements of persons" tiny and invisible.[28]

Latimer's political vision emerges from a conception of coral and coral islands that differs markedly from, and is more scientifically accurate than, that which informs the work of Beecher and Spofford. For Latimer imports a central insight of Darwin's 1842 treatise on coral, which was examined in the previous chapter: reefs are always in the process of becoming something other than their past and present forms could ever allow us to predict. Within a Darwinian conception of coral, "the relations . . . which determine the formation of reefs on any shore . . . must be very complex, and with our imperfect knowledge quite inexplicable."[29] The smallest movement of a single entity could dramatically alter the entire structure over time.

In the indeterminacy and potentially transformational relations that Darwinian reefs materially embody, Latimer and other nineteenth-century US women found hope for a different human future that could emerge from the wider uptake of the work women were already performing. Because the tiniest "movements of persons" "never known" may create a "world changed for all time" (in Latimer's words), women have every reason to continue to hope and to work toward "new conditions of existence." Rather than declaring that our "society is that island," along with Beecher, or assuring readers that we are all already "doing something for someone else, whether physically, spiritually, or intellectually," along with Spofford, Latimer exhorts us to action: "assume the relations."[30] Her reef politics do not tell people to accept their present unjust conditions in exchange for a future, enormous, abstract result that their sacrifices will inevitably bring about. "We, the corals" acknowledges that, presently, the body is the shaping condition of one's existence—the condition that determines which modes of social and political action one can perform and what kinds of support and recognition one receives—while also signaling that it need not be so, that the point is to build something new. Reefs tell us that the strongest and most robust polities emerge by cultivating interdependence, which involves acknowledging, and continually contesting, the ongoing ways that bodily features stratify persons socially and politically.

The idea that a small "insignificant" being may dramatically alter the entire structure by forming "relations" with others is a defining claim of reflections on coral and politics by a number of Latimer's contemporaries. According to US feminist philosopher Ellen M. Mitchell, humans would ideally operate according to Darwinian reef principles.[31] In "Corals" (1873), an essay appearing in the popular women's periodical *Arthur's Illustrated Home Magazine*, Mitchell analogizes "corals" (by which she means coral polyps) to women's bodies as more conventionally imagined (or possibly even to her own body), writing that corals are so physically diminutive that "one can hardly perceive them," they have "red lips" and are "very sensitive," and when we examine them closely, "we find the almost unknown body of this stone animal as fearfully and wonderfully made as our own."[32] But the core of her analogy of coral and women is that the labors performed by corals are like those society-sustaining human labors that, if more widely taken up, might just render the polity as mutually sustaining as a reef. For coral labor involves continually "moving, feeding, [and] producing others" before being "turned into stone and buried in its own rocky house, while countless generations build new abodes on its grave, and in their turn pass away and are followed by others."[33] That work of binding is embattled and multigenerational: "Year after year, generation after generation, the small and lowly polypi work at their heaven-appointed task, in quiet and silence, with modest industry and untiring energy; the tempest beats upon their fragile homes, and the mighty waves thunder against them . . . but the living force, though so small, conquers at last."[34] Ultimately such labor can yield a better society. The coral, she writes, is "a perfect socialist and communist" in its "common labor" and capacity to "absorb" and deposit resources "unceasingly," for "after taking the first taste" of food, each "sends the surplus down into the common receptacle, from whence it is afterward distributed equally through countless channels into all parts of the [coral] tree."[35] Unlike Beecher's imagination that sustenance in a reef naturally "flows" in equal distribution to all, Mitchell clarifies that only hard work makes that happen.

And thus we humans should all be like coral and try to build a reef. For, Mitchell concludes, reefs tell us that "the humbler are the more useful, the smaller the more powerful. Would not man do well to ponder this truth, when tempted to complain that his sphere of action is limited, his means for doing good circumscribed?"[36] Here Mitchell echoes Latimer in urging "man" toward a particular kind of "action" that involves daily, ceaseless, local, and less historically visible labors, for these have produced "glorious results." The implication is that we need not wait to perform or benefit from some grand and public gesture; our best chance of achieving that equal distribution and mutual sustenance so admirable in the coral ("a perfect socialist and communist") is for each of us to direct "modest industry and untiring energy" to that "moving, feeding, producing others" in which corals already engage. "Even the [coral] jewels we wear are eloquent with beautiful teachings" of that lesson, which is ultimately Mitchell's direction for us to do what Latimer states more explicitly: "assume the relations" and know that the system can be "changed for all time" by the tiniest "movements" of a single being so "small" as to be "invisible."[37]

That lesson is given narrative form in an early and relatively neglected short story by Harriet Beecher Stowe called "The Coral Ring" (1843). A tale of temperance that has generated almost no scholarly commentary, "The Coral Ring" seems at first to have nothing at all to do with coral. The tale's eponymous "coral ring" receives a single mention in the story, and the word "coral" appears exactly once.[38] But coral is more than incidental to this seemingly forgettable early work by a writer who would go on to pen the most popular abolitionist novel of the nineteenth century.

"The Coral Ring" teaches upper class white women how to be like coral and build a reef by forging and sustaining nonbiological bonds, a form of binding that men should value, support, and participate in. The story centers on the relationship between the wealthy, educated Florence Elmore, age twenty, and Edward Ashton, her "old bachelor cousin" who awakens Florence to her transformative potential to alter society by forging sustaining bonds among those nearest to her, beginning with those most in need—in this case, Colonel George Elliot, whose intemperance has made him a social outcast beyond the help of any "*man* of his acquaintance."[39] At the start of the story Florence is frivolous—like most girls of her age, according to Edward. Where "among all our circle," he asks Florence, will you find a young girl "that has any serious regard for the improvement and best welfare of those with whom she is connected at all, or that modifies her conduct in the least with reference to it"—present company *not* excepted?[40] Florence protests half-heartedly, then agrees that her life lacks "purpose," and that young women of her class generally are "fit for nothing on earth but parlor ornaments."[41] While she admits that she has "often had a sort of vague idea of something higher that we might become," she fails to imagine how to contribute other than by being "amiable in the family." She asks, "What more than this is expected of us? What else can we do?"[42] To this Edward replies with another question: "Are you not responsible . . . for the influence you exert over those by whom you are surrounded?"[43] At this point Edward unfolds the sordid tale of Colonel Elliot, "a lost man" steadily abandoning his "brilliant talents" and social success to "intemperate habits," and encourages Florence to "save him."[44] After significant initial resistance, Florence concludes, "I will try."[45]

The remainder of the story focuses on Florence's efforts to reach out to Elliot, build on her "platonic sort of friendship" with him, and bring him back into the society he has fallen out of, efforts that Stowe describes as a type of transformative, interpersonal binding. One day, as Florence and Elliot read together, he notices her ring, which features "a coral cross set in the gold." It's a "knight's ring," she explains, worn by "the red-crossed knights." The red cross of the medieval Knights Templar, said to have abstained from alcohol, was adopted by several nineteenth-century temperance organizations, and many even called themselves "templars." But here I am less interested in this meaning of Florence's ring than in what *Florence* makes of the coral object: "Come, now, I've a great mind to *bind* you to my service with it," she tells Elliot, who swears obedience, wearing the ring as a pledge.[46] Later, when the two attend a party together, Florence prevents Elliot from drinking by merely pointing to the ring and then encouraging him to sign a temperance pledge not only for his own sake but also "for those whose life I know is *bound* up in you."[47] Elliot obeys and credits Florence with doing "what, perhaps, no one else could have done," words that bear out Edward's initial claim that women such as Florence are "responsible" for social "welfare" due to their potential to adapt to, lift up, and sustain those with whom they are connected, both biologically *and* nonbiologically.[48] As Stowe's narrator observes, "the benefit of the whole" would be encouraged if everyone learned to act as Florence has.[49]

FIG. 4.1. "The Coral Ring": The title of Harriet Beecher Stowe's 1843 short story may be a reference to Darwin's recent description of a coral atoll, or "ring-formed reef," which is expanded across time by tiny polyps. Darwin's sketch of an atoll in *The Structure and Distribution of Coral Reefs*, 1842.

"The Coral Ring," then, is about what Latimer might call the "new conditions" that we can imagine and work toward—even, and perhaps especially, from a somewhat limited social position—when we "assume the relations devolving upon us with modesty and discretion, yet with energy and boldness."[50] Florence learns to act like the coral she wears, a small body ever expanding beyond the biological family to "bind" and sustain, and her life should inspire other white women of a certain class to do the same. This 1843 story asks readers to subscribe to a central tenet of coral reef formation elaborated by Darwin in 1842: the unexpected, unpredictable, and tiny movements of small bodies may produce the most transformative alterations in the whole. Considering the story's emphasis on the upward potential of small bodies binding—as Florence herself says, there is "something higher that we *might* become"—one wonders whether "The Coral Ring" plays on the name of a geological formation that Darwin describes in his treatise: the coral atoll or "ring-formed reef" (fig. 4.1).[51]

To be clear, Stowe is emphatically not suggesting that society is a Darwinian "coral ring" but rather that it "might become" one if more of us enacted a laborious binding—in this case, social reform—that sustains others in need. And reform was not the only form of nonbiological binding through which nineteenth-century US white women and men could be like coral.

Anstrice Fellows (1800–75) and Eunice C. Fellows (1802–83), sisters from Ipswich, Massachusetts, turned to the process of reef formation to conceptualize and promote their life's work of placing orphaned children with adoptive parents, a project that they hoped would transform society by redressing present class disparities. The sisters ran a free adoption service in Boston that required families to treat children "as equal members of the family circle," an endeavor that they supported by editing their journal, *The Orphans' Advocate and Social Monitor*.[52] The journal's May 1850 issue features a remarkable short essay on coral that analogizes adoption to the work of reef-building through "Small Efforts well Directed" (the essay's title) that could ultimately produce a more equitable society:

> There are no builders in the world like the coral. This little insect can do what the monsters of the deep, and the monsters of the land, and the skill and toil of man cannot accomplish. . . . May we not learn a lesson of the coral? No splendid efforts which orators put forth, nor powers of legislation are competent to transform the character of a nation. But mothers, exerting a constant, unregarded influence over the minds of the children, may make the next generation entirely and dramatically different, either for the better or worse, from what the past has been.[53]

According to this passage, the "lesson of the coral" is the very Darwinian claim that "constant" labors of "little" bodies "may" ultimately "transform the character of a nation" into something "dramatically different" from its "past" forms. But how, exactly, and in what way?

The remainder of "Small Efforts well Directed" makes the case that the ongoing labor of forging nonbiological bonds through adoption is perhaps the most effective way to redress the inequities of our present, and thereby build a more just future—though currently such labor falls to persons "silent and unobserved." To support this assessment, the sisters turn to the example of prison reform. Yes, this "noisy benevolence" does some good yet not nearly as much as "the taking of the little children from the midst of peril and temptation, from which, in the natural order of events, they would be hurried on into crime, and then into prison." If more people were able to "trace the influence of their conduct to the end," then they would "act more wisely, and accomplish more good, than they could do by merely casting their gifts upon the current of benevolence, which flows with noise and observation, beside them." The essay ultimately fashions adoption as a laborious, silent, yet crucially constitutive, act of what Latimer might call "assuming the relations": for only by "taking" in and on the "responsibility"—not by "casting" one's "gifts" "beside" one—can we possibly hope to "transform" the nation into something new and better. And though presently that labor falls to women, the essay hopes that all "those who desire to do good" will reconsider their options by considering the "lesson of the coral."[54]

## AFRICAN AMERICAN CORAL COLLECTIVES

Thus far we have been examining the work of white women for whom coral illuminated existing gender and class inequities and encouraged more people to take up labors that could produce a structure less fully reliant on extractive labor relations that constrained and marginalized some groups while disproportionately enriching others. Yet what about racial inequities? Put otherwise, what kind of racial collectives were these white women imagining? For these women coral's power to encourage collective binding quite likely stopped short at race, which is one important reason that Black writers took up and adapted coral to their own political ends across the entire second half of the nineteenth century.

During the violently divisive decades of Reconstruction, Frances Ellen Watkins Harper chose coral as an analogy to promote the inter- and cross-racial bonds that she considered central to the success of that political project. Here is Harper's poem on coral and politics, "The Little Builders," which she recited to Black and white audiences on her lecture tours of the US South between 1864 and 1871 and then published in her 1871 poetry anthology intended for an even wider audience:

> Ye are builders little builders,
> Not with mortar, brick and stone,
> But your work is far more glorious—
> Ye are building freedom's throne.
>
> Where the ocean never slumbers
> Works the coral 'neath the spray,
> By and by a reef or island
> Rears its head to greet the day
>
> Then the balmy rains and sunshine
> Scatter treasures o'er the soil,
> 'Till a place for human footprints,
> Crown the little builder's toil.
>
> When the stately ships sweep o'er them,
> Cresting all the sea with foam,
> Little think these patient toilers,
> They are building man a home.

Do you ask me, precious children,
How your little hands can build,
That you love the name of freedom,
But your fingers are unskilled?

Not on thrones or in proud temples,
Does fair freedom seek her rest;
No, her chosen habitations,
Are the hearts that love her best.

Would you gain the highest freedom?
Live for God and man alone,
Then each heart in freedom's temple,
Will be like a living stone.

Fill your minds with useful knowledge,
Learn to love the true and right;
Thus you'll build the throne of freedom,
On a pedestal of light.[55]

The poem's speaker directly addresses emancipated Black children and the Black women who guide Black children's "little hands" toward "useful knowledge," as literary scholar Karen Chandler observes.[56] The goal is to build "freedom's temple" in the specific way that polyps build reefs. And the extended analogy of Black children and women to coral builders encourages these recently freed people to participate in "creating the conditions of their own emergence into full citizenship," observes literary scholar Mary Loeffelholz, for the poem fashions the self-directed education of Black persons as a mode of "building" that promotes the "emergence of a new kind of national body, not yet fully visible"—a "future yet to be."[57]

I am deeply indebted to Loeffelholz's (2004) extensive analysis of this poem, which in turn builds on formative studies of Harper's work by Frances Smith Foster (1990; 1993) and Carla Peterson (1995).[58] In what follows I extend some of Loeffelholz's claims by attending to Harper's deep engagement with the materiality of coral, while also situating Harper within a longer tradition of reflections on coral and politics that engaged some of her white contemporaries. For no scholar has yet observed that Harper's "The Little Builders" is deeply in dialogue with a much earlier work, a hymn on coral and politics of the same title that was first published in 1846 in an evangelical periodical for children produced by the London Missionary Society.[59]

Harper's revisions of this source material dramatically emphasize her perception of coral's power to galvanize nineteenth-century US Black women and white people in a project of mutually transformative, sustaining bond building that she believed Reconstruction required. Harper might easily have read the evangelical "Little Builders," which was reprinted in several US periodicals, or perhaps she sang it.[60] Either way, her poem shares more than a title with the earlier work. It also shares some vocabulary, including the phrase "little hands," and both pieces describe children and corals as "little builders" whose labor could gradually produce a massive and socially beneficial result.

Yet here Harper's poem diverges quite sharply. The goal of the original "Little Builders" was to galvanize children into a more active role in the British empire's missionary project in the South Pacific.[61] The poem encourages "little hands" to bring "little offerings" that fund the building of "houses" in "heathen lands" and make way for metaphorical "dwellings" for the "holy word" in heathens' "hearts like mountain stone." "Little Builders," then, interprets coral as a natural model of a familiar and widely adopted form of human colonial expansion in which white persons and their cultural practices spread ever outward, gaining strength and stability by displacing Indigenous peoples and cultures, building over their land, and even invading their bodies.

By contrast, Harper's reef expands by sustaining others, a process that must begin by centering the racially disfranchised. It also requires the continual labor and adaptation of all involved and hopefully produces a polity that is different from and more inclusive than the present one. From the start we learn that Harper's "little builders" are not building literal buildings at all. Unlike the builders of the earlier poem, these ones build "not with mortar, brick and stone." And even the metaphorical "buildings" they construct differ drastically from the "dwellings" for the "holy word" that the earlier poem's builders establish in heathens' "hearts like mountain stone." The "buildings" of Harper's poem are the people themselves—just as corals are both the creators of the reef and also of the reef itself. When her speaker tells listeners and readers to make their "heart . . . a living stone" by continually gaining and giving "useful knowledge," she means that each person should enact and experience a transformative binding. Or as Harper elsewhere told listeners, Reconstruction should "not consist in increasing the privileges of one class and curtailing the rights of the other, but in getting every citizen interested in the welfare, progress and durability of the state."[62] By contrast with the "mountain stone" of the earlier poem, Harper's "living stone" is processual; it is, as Loeffelholz observes, ever gradually becoming "a new kind of national body."[63]

The coral of Harper's poem thus brings together and affirms what Foster identifies as the two central goals of the Reconstruction lectures through which Harper sought "to bind the nation's wounds."[64] First, Harper held that a more equitable, mutually sustaining future for the nation "depended upon the ability of its citizens to unite behind a common goal," to establish a true "community of interests" as Harper called it.[65] Second, Harper imagined Black women participating centrally in that process, while she also demanded that whites commit energy and resources, too.[66] For while the poem addresses emancipated Black children and women, it widened its circle of accountability as it moved from Harper's lectures into print, finally addressing "the nation itself," which is the intended audience of Harper's 1871 anthology containing "The Little Builders."[67]

There could scarcely be better evidence of coral's historic power to exhort marginalized persons to work from constrained conditions, and always toward a more mutually sustaining society, *while also* calling on everyone else to share in that labor. To those many white women who asserted that reefs grow most robustly by sustaining a vast variety of others, rather than exploiting and displacing the many for the few, Harper seems to be pointing out the exclusions that continue to haunt even that reef romance. Put another way, her poem seems to be redefining the "we" of Latimer's "We, the corals" to center Black women while also calling on all readers to perceive that recently freed Black persons are those most in need of resources.

In that interpretation of reefs, Harper participated in a tradition of Black writing and thinking about coral and politics that began in the 1850s with James McCune Smith and persisted until at least the turn of the twentieth century. During this period a number of Black writers, speakers, and activists drew on the relational nature and growth of coral to give voice and authority to some of the principles of civic identity and participation that literary scholar Derrick Spires identifies in the earlier work of Smith, Harper, and other Black writers, and that constitute an alternative to "the citizenship from which black Americans had been and continue to be excluded."[68] This alternative form of citizenship, Spires shows, required that each person "create the bonds and material support through which the collective could make larger scale changes."[69] It involved not individual labor for private accumulation but rather an "active engagement in the process of creating and maintaining collectivity, whether defined as state, community, or other affiliative structure."[70] And crucially, this form of citizenship is emergent: rather than a destination—a concretized identity or fixed relation to a state—it is "a self-reflexive, dialectical process of becoming."[71] In this context, reefs abetted Black political theorizing by materially modeling the particular claim that a single and seemingly insignificant entity, by way of the relations it creates and sustains with other entities past and present, might transform the entire system into something new and different.

Black writers and theorists took up and adapted that central insight to different political purposes at different historical moments across the tumultuous and violent decades between 1859 and 1900.

In the pages of the *Anglo-African Magazine*, for example, Smith published an essay entitled "Civilization" (1859) in which he draws on coral to argue that a diverse and thriving society could emerge only from the attenuation of "caste and slavery," for slavery and "the iron walls of caste" are the only remaining barriers "in the way of our national advancement."[72] Smith grounds that argument in an elaborate reef metaphor that draws the essay to a close: "Away down in the depths of the ocean . . . the coral insect toils on through years and years; the insect perishes, but its labors live, and pile on pile, its tiny successors continually lay" until they "reach the surface of the sea" where they are rewarded for their labor in the form of an island that supports all of society. Smith goes on to say that the work of Black abolition is similar to yet "higher than the labor of the coral," for on the ongoing project of abolition depends "the progress of mankind."[73] As scholar of American literature and culture Juliana Chow observes, here and in other writings Smith's argument for racial uplift finds "force and vitality" in scientific accounts of coral by Darwin and others who describe reef formation as an imperiled, immense, slow, unacknowledged, discontinuous, yet ultimately indispensable labor.[74] But Smith also grounds his argument in a very particular idea that Darwin elaborated first and most fully in the coral treatise (1842): all emerges continually from relations among multiple and diverse entities, relations enabled by seemingly small beings who could thereby produce a dramatic alteration of the structure over time.

In this essay Smith does not explain precisely how Black communities can start to achieve that transformational assemblage in daily practice, but an essay published in the same journal later that year does, and the writer partly rests his case on principles of coral's growth. Possibly in answer to Smith's argument that slavery and "the iron walls of caste" are the only remaining barriers "in the way of our national advancement," Black abolitionist and journalist Jonas Holland Townsend (1820–72) writes the following: "The mighty agencies to be employed in battling down the strongholds of caste and prejudice, which now operate so powerfully against us, are to be found in the more perfect development of our intellectual powers and capacities" through educating Black youth in common schools and colleges.[75] For our potential "future greatness," Townsend argues, will derive not from the individual pursuit of wealth, which makes us "servile tools of the white man," but rather from pursuing labors that sustain "the millions with whom we are identified." Black education will "stem the tide" of prejudice by shaping "all our general relations of life"—just as "the coral insect, by its continued labor, builds a reef that stems the mighty currents of the ocean."[76] For Townsend, reefs offer an alternative to capitalist individualism, which will only bring more of the same; coral insects, by contrast, show us that "revolution" depends on actively fostering and sustaining new and various "relations of life."[77]

Decades later, in the wake of Reconstruction's failures, coral insects gave Black civil rights activist Ferdinand Lee Barnett a powerful way to remind Black Americans to cultivate economic interdependence as the basis of collective material security and thriving. In a speech at the 1879 National Conference of Colored Men, Barnett—Chicago's delegate to the Convention and later the husband of Ida B. Wells—argues that Black social and political advancement would come not from the pursuit of "individual brilliancy" or "the preferment won by a few wealthy ones" but rather from "labor for the benefit of the multitude," guided by "fraternal feeling" for others past and present.[78] Barnett authorized his claim with reference to coral, telling listeners the following:

> Many thousand years ago, a tiny coral began a reef upon the ocean's bed. Years passed and others came. Their fortunes were united and the structure grew. Generations came and went, and corals by the million came, lived, and died, each adding his mite to the work, till at last the waters of the grand old ocean broke in ripples around their tireless heads, and now, as the traveler gazes upon the reef, hundreds of miles in extent, he can faintly realize what great results will follow united action.[79]

In sharp contrast to Beecher's "waves of wealth" that augment "society" as individuals labor for their own accumulation (and sometimes "fail" and become engulfed and forgotten in the process), Barnett's island arises only from every individual actively distributing the surplus through a carefully managed, "rigid economy" that sustains everyone involved.[80] In this model of a reef, the builders are also the beneficiaries since their "tireless heads" eventually rise above the waves.[81]

And in "The Negro Element in American Life" (1900), an important yet relatively understudied oration by African American minister Abraham Lincoln DeMond, a reef attests that building a more just present and future for the country as a whole must begin by recognizing that Black people *have always been* fully constitutive of the society from which they and their contributions have been systematically erased.[82] Speaking on the first day of the twentieth century at the Dexter Avenue Baptist Church in Montgomery, Alabama—the same church from which Martin Luther King Jr. would lead the Civil Rights

Movement—DeMond used coral to launch what we might productively imagine as his own *1619 Project*. DeMond declared,

> The Negro's story must also be told, for the history of the nation will not be complete without it. Humble as may have been his offering at times, and, though given in tears and blood, it has entered into the structure of this fair Republic of the Western World.

> "The coral insect but an atom gave
> To help uprear the pile he ne'er could see;
> But now it towers above the topmost wave,
> He has a part in mansions yet to be."

Directly following this poem—which appears to be DeMond's own creation—DeMond draws the following lesson from coral: "The Negro has a part in the history of this country of which he need not be ashamed. Let it be told for his own vindication. Let it be told as an answer to those who would slander the race. Let it be told for the encouragement of the rising generation. Let it be told for the sake of truth and eternal justice."[83]

Of vital importance to DeMond's interpretation of coral is that coral insects are not dead and gone and forgotten: reefs only emerge because coral builders have always "entered into the structure" and each "has a part in" its future. That conception of a reef, in turn, gives form and authority to what DeMond told his listeners next: "The Negro element in American life made its first appearance at Jamestown in the year 1619," when Black men arrived to labor, thereby saving the colony from extinction.[84] That moment is, DeMond explains, the beginning of the systematic enslavement of persons of African descent through which ultimately "the race became a source of revenue and all the States profited by it," as the labor of Black persons transformed the North American wilderness into civilization. Though "a silent factor, he was a factor still" and "can not be eliminated from the history of that time. He does the race an injustice who fails to recognize what it has contributed in toil for the upbuilding of this nation."[85]

DeMond's speech makes an especially fitting conclusion to this chapter because, through a coral reef metaphor, he identifies and seeks to amend the very elision of Black lives, knowledge, skills, and labor that so many white writers furthered via the same metaphor, as I demonstrate in chapter 2. Those writers, too, held that coral insects drive human history by making and shaping its material foundations, and one white writer in particular credits coral insects with making the US into the society as we know it, beginning with the construction of Plymouth in 1620: Jane Andrews's short story, "Sea-Life" (1866), celebrates the creation of a coral island called "Coraltown" by "emigrant" coral insects, just as "the Pilgrims at Plymouth" are the true founders of Massachusetts.[86] The elaborate reef metaphor that Andrews develops across "Sea-Life" ultimately indicates that the most foundational, essential, constitutive, self-sacrificing laborers—those without whom we could not have the society we now know—are the white Europeans who landed at Plymouth in 1620. In this way the narrative elides other stories of nonwhite life and labor, such as those of Black and Indigenous persons.

The enormous, and still astonishingly relevant, risks and stakes of DeMond's reef metaphor emerge all the more clearly when considered alongside Andrews's coral story of American origins.[87] By using coral to illuminate the centrality of Black lives and labor to the country's formation and expansion, DeMond speaks back, demanding recognition of the spatial, material, and spiritual presence and significance of persons of African descent through the very metaphor that had historically opposed that project. It is almost as though DeMond is telling white writers that they could not have coral, because what a reef most keenly renders for humans is the exploitation, dispossession, reduction, and erasure of African descended persons from the republic that they laboriously formed and continued to sustain, and also because reefs model the lesson that any truly collective present or future "rising" depends on assuming relations to the past that are at once more implicated and historically accurate.

INTERCHAPTER

# "The Coral Builders," St. John's AME Church, Cleveland, Ohio

In late nineteenth-century Cleveland a Black literary society called "The Coral Builders" put into action a number of the relational and transformative bond-building practices that Black writers and activists, from Smith to Harper and Barnett, theorized and encouraged through coral. Founded in 1892 by a group of young members of St. John's African Methodist Episcopal (AME) Church, and guided by their first President, Frank Lee, a civil rights advocate and community leader who had been born into slavery in Virginia, the Coral Builders Society, like other Black literary societies of the period, created a forum for Black civic participation, economic uplift, joy, and social connection for almost a decade.[1] We may never know what compelled the group to name themselves after reef-building corals at a Society meeting in late November 1892, just a month after their founding. Interestingly enough, in the late nineteenth-century US it was not unusual for small groups of white parishioners, usually women, to call themselves "Coral Builders" or "Coral Workers" after the relatively small and incremental funds they raised through sewing, cooking, and other activities to support their church or an overseas mission.[2] The record of St. John AME's Coral Builders, however, attests that for nearly ten years this group worked dynamically and responsively, and always toward a more mutually sustaining polity, by creating and cultivating a variety of bonds, past and present, both within and beyond St. John's congregation, while calling on an ever-expanding range of others to do the same.

When St. John's AME Church (which stands to this day) was founded in 1830 by Reverend William Paul Quinn and a small group of formerly enslaved people as the city's first Black church, only seventy-six Black people lived in Cleveland.[3] Over the ensuing decades Cleveland's Black population grew significantly, while the city became an important site of abolitionist activity and a crucial stop on the underground railroad, as both Black and white antislavery activists spearheaded and achieved legal and political changes of national significance.[4] By comparison with other Northern urban places in the middle decades of the nineteenth century, Cleveland offered a measure of racial tolerance; yet that tolerance began declining after the Civil War, as legal segregation and discrimination spread during and after Reconstruction (1865–77), even while Cleveland's Black population doubled in size by the 1890s due to migrations from the South.[5] Though St. John's AME had struggled financially since its founding in 1830, the church sustained a steadily growing membership through services, support of civil rights, and social outreach. During the 1890s that membership comprised mostly working- and middle-class Black people, including many migrants from the South, and St. John's was meeting regularly in its own building on Erie Street (now E. 9th Street), which members sometimes called the "Old Erie Street Church."[6] Moreover, it was one of three Black churches and part of a thriving cultural scene that drew writers, musicians, and activists whose work was broadcast weekly in the city's Black newspaper, the *Cleveland Gazette*, founded in 1883.

From October 1892 through April 1901 the *Gazette*'s third-page column on local events carried regular and lively announcements of the meetings and other activities of the Coral Builders Society, begun "by a large number of young people . . . to improve their literary and musical talent, as well as to assist in defraying the gas expenses of St. John's church."[7] Over the next ten years, a period of particular Church activity and fundraising, the Coral Builders planned and carried out an astonishing array of events that raised enough money to cover much more than the Church's gas bill, while achieving significantly more social good than the modest goal of member self-improvement.

The society usually held its meetings on Wednesday evenings in members' homes, and these gatherings would often begin with an address to the group by one of its members or a guest speaker and then proceed to society business, including elections for positions ranging from president to vice president and executive committee and planning for events hosted in homes, at the Church, and in public venues that sometimes drew large crowds. By December of 1892 membership numbered fifty and was on the rise, and by July 1893 the Society was in "splendid condition financially" and able "to assist in paying off the main debt of the church."[8] During the following year they began "raising funds to pay for the vocalion," a form of reed organ that was popular during the last decades of the nineteenth century.[9]

The society's events indicate an interest in cultivating the collective as the necessary foundation of Black social and economic uplift.[10] According to notices in the *Cleveland Gazette* the society both celebrated milestones in the specific lives of its members—through wedding receptions and birthday parties—and brought together members and nonmembers alike through events including socials, literary and musical programs, lectures, readings, concerts, plays, fairs, soirées, dinners ("oyster suppers" were a favorite), lawn fetes, a camp meeting, and a variety of fund-raising activities that have since fallen out of fashion and speak to the energy and ingenuity through which the society created social and economic bonds. These include a "cane drill," or synchronized marching routine, performed by a group of thirteen women belonging to the society's "ladies drill corps"; a "'Martha Washington' entertainment" replete with colonial costumes and possibly songs and speeches; a "dumb" social where fines were levied of anyone who spoke, whispered, laughed, or otherwise verbally communicated during a set period of silence; and something called a "bank opening," in which contestants evidently allowed the sum in their bank account at the recently opened Dime Savings and Banking to be recorded and read aloud to the community, after which they then donated to St. John's.

While Society members themselves frequently hosted and performed in these events—members put on a great variety of musical performances, comic plays and dialogues, and "addresses" on various topics—they also sponsored notable out-of-town guests, such as the Reverend J. M. Henderson and a pianist from Oberlin's conservatory. A number of the society's events attracted especially large audiences, well beyond the congregation of St. John's and sometimes to other venues around the city. The members' performance of a farcical play called *The District School at Blueberry Corners* (1893) was a particular "artistic and financial success," eliciting rounds of applause and demand for an encore from the audience gathered at St. John's. The cane drill, "executed almost faultlessly," drew a crowd to Excelsior Hall where they were treated to a variety of other performances including an operetta and a reading. And "Master Royden Cook, the boy elocutionist" was a particular hit.

The society's activities indicate a commitment to putting the collective first, to making the collective, including a more collective distribution of surplus value, the necessary foundation of individual thriving, rather than individual labor for the accumulation of individual wealth. Their approach to uplift may have been modeled on the remarkable life and actions of their first President, Frank Lee.[11] Lee was about fifty when he helped found the Coral Builders in 1892, by which time he was married and living in his own home on Webster Street as a pillar of the church community, an important civic figure, and an outspoken political activist for the Republican party. These achievements are especially notable given the circumstances of Lee's early life. As he documents in his testimony for a 1904 pension application, he "was born in slavery in Eastern Virginia" and was uncertain of his age because his "folks were illiterate and kept no family record."[12] Around age twenty, Lee "escaped from [his] master and followed the union Army" north, then enlisted in the Fifth Massachusetts Colored Cavalry Regiment. Army life took Lee back to Virginia, where his regiment fought in the Siege of Petersburg and later numbered among the first Union forces to enter Richmond triumphantly once the Confederacy abandoned the city in April 1865.[13] After the War, Lee eventually settled in Cleveland, working first as a janitor then as superintendent of St. John's AME Sunday School.

By all accounts Lee's life in Cleveland was dedicated to many of the ideals of Black citizenship that Spires elucidates among earlier nineteenth-century Black communities for whom citizenship was an ongoing and responsive practice of forging more just bonds temporally and spatially, past and present, social and economic, which could become the foundation of larger scale transformations.[14] By the mid-1880s Lee was active in several churches and regularly presenting speeches and addresses at a number of Black literary societies. He joined the Grand Army of the Republic, a patriotic organization of Civil War veterans, and worked particularly to keep alive the memory of African American soldiers as a way of inspiring the present generation in their fight against segregation.[15] In the wildly successful bank opening hosted by the Coral Builders in March 1900, Lee came in third, which means that the sum in his bank account made up a significant proportion of the funds that defrayed the church's debt for the vocalion. In response to the murders of African Americans by local white supremacists in Wilmington, North Carolina, in 1898, Lee joined with local religious leaders to found the short-lived "Brotherhood of African Descent," which he envisioned as "a national organization of colored citizens" dedicated to "united political action."[16] And when Lee died in 1907, he left a will placing most of his property in a trust divided between Cleveland's Home for the Aged Colored People and St. John's AME.

Viewed in light of these actions, the Coral Builders seems a quite direct outgrowth of Lee's approach to collective uplift. As a group of individuals collaboratively fostering the intellectual, economic, cultural, and social development of Black Clevelanders for nearly a decade, the Coral Builders more than bore out the principles of relational and transformational thriving that Smith, Harper, Townsend, Barnett, DeMond, and surely other contemporary Black writers and speakers attributed to coral. The society's history attests that Black communities successfully harnessed the material features and imagery of reefs for their own political ends, using coral to imagine a "we" that is fully founded in relations, past, present, and future.

# Red Coral, Black Atlantic

During the second half of the nineteenth century, US fiction about slavery and race often featured descriptions of objects made of Mediterranean red coral. We can probably never know how much the authors of these texts knew about the history of coral as currency in the transatlantic slave trade, the Indigenous African value systems that made red coral valuable enough to accept in exchange for persons, or the persistence of those value systems throughout African diaspora communities in the Atlantic world, from Jamaica to New Orleans, where Black women chose to wear red coral as a marker of social status and personal and collective identity. But in this chapter I examine these histories and uses of red coral as they come to bear upon US fiction, and moreover on conceptions of Black life and identity in the nineteenth-century US. Turning from visions of coral collectives to the material culture of coral objects, I explore five major works of US literature in which items made of coral—necklaces, earrings, sleeve buttons, and pins—drive the story because of what Black women know about them.[1]

The texts in question are, in chronological order, *Uncle Tom's Cabin* (1852), a popular abolitionist novel by Harriet Beecher Stowe, a white Northerner; *Peculiar: A Tale of the Great Transition* (1864), a rambling and almost picaresque postemancipation novel by Epes Sargent, a white Northerner, abolitionist, editor, and textbook author; *The Grandissimes: A Story of Creole Life* (1880), a historical novel by George Washington Cable, a white Southerner, that tells of racially diverse New Orleans on the cusp of Louisiana's US annexation in 1803; *Pudd'nhead Wilson* (1893), a historical novel by Mark Twain, a white Missouri-born writer, about racial tensions in small-town, 1830s slaveholding Missouri; and "Her Virginia Mammy" (1899), a short story of Reconstruction-era race relations in a fictional town strongly resembling Cleveland, Ohio, by Charles W. Chesnutt, a Black Northerner.

All of these texts, along with many other US fictions about slavery and race from the period in question, have been the subject of much scholarship. By and large scholarly inquiry has focused on how such texts engage the nineteenth-century US definition of race that rationalized slavery. According to an indispensable scholarly consensus, these various texts centrally grapple with the ideology that race is a biological fact, a perceptible feature of the body, and a reliable index of what is more difficult to see—such as personal character and intellectual capacities—whether to support, challenge, contest, revise, reject, dramatize, or otherwise explore that ideology.

While that is true, one risk of repeatedly reading these works largely in relation to the account of race that licensed (and licenses) so much racial violence—even if our goal is to critique that account—is, as Katherine McKittrick argues, that we end up granting it even more space and power to define and contain Black life, past and present. For McKittrick, the goal should be rather to "think about the social construction of race in terms that notice how the condition of being black is knotted to scientific racism but not wholly defined by it."[2] In that spirit, the present chapter seeks to "notice" the framework of scientific racism, as it shapes all of these narratives and particularly their depictions of Black women, while also noticing that the red coral ornaments that Black women repeatedly reach for and wear, evoke and describe, for purposes that matter to them, also point us beyond the spatial and temporal boundaries of the US altogether, and toward what Derrick Spires might call "Black expressive culture that exceeded US cultural, political, and physical borders."[3]

Generations of Black women across African diaspora communities in the Atlantic world during the eighteenth and nineteenth centuries wore red coral ornaments, particularly on special occasions, to evoke a shared African past and create a diasporic identity and community that was forged by slavery yet not wholly defined by it.[4] It is not possible to know precisely what red coral meant to these women or precisely how that meaning changed among different groups of these women over time. We can say, however, that to them red coral figured, or figured importantly within, what McKittrick calls "black Atlantic livingness," ways of being human, and of being in relation to one another, that can neither be expressed nor understood within a colonial framework because they precede and exceed colonialism's hold.[5] What appears, through red coral, in the nineteenth-century works of fiction under consideration in this chapter, then, are fragments of a Black Atlantic ritual that precedes and exceeds not only US boundaries but quite likely the full understanding of the authors and most readers.

Some explanation of my method is in order, for how could numerous works of fiction about race and slavery, most of them by white US writers, collectively register and transmit an understanding of red coral that the writers themselves did not, in all likelihood, comprehend? For one thing, this is partly how "scriptive things" work: as Robin Bernstein explains, certain items of material culture may, as a result of their histories of meaning and use, generate a "script," defined as a shared set of actions or ideas that repeats with variation across time and contexts, and whose full meaning may "include and exceed" the intent of any particular actor or author, even as that author records the script in their work.[6] Black women passed down knowledge across generations, both through stories and everyday practices, and so they could also pass along the meaning and importance of red coral to one another. White writers living across the US need not have understood what red coral meant to Black women in order to notice its significance to these women and to write about that significance.[7] Considered in this way, it makes sense to approach these widely read works of US fiction as we might approach earlier ethnographic texts produced by white observers of Black populations in other parts of the world: these texts are not documentary, for the subjects portrayed in them had little to no control over their representation. But these texts can become useful sources of knowledge about the subjects they portray when we read them with caution and corroborate them with reference to one another and to other types of sources, such as oral traditions, archaeological evidence, and histories of Afro-Atlantic customs and performances.[8]

Accordingly, the first part of the chapter focuses on a red coral necklace worn by the character of Palmyre Philosophe, free person of color and New Orleans Creole woman of Senegalese descent in George Washington Cable's *The Grandissimes*. I take that necklace as provocation to trace a necessarily fragmentary genealogy of Black Atlantic knowledge of red coral ornaments. The second and third parts of this chapter interpret US fictions of race and slavery in relation to that genealogy. Epes Sargent's *Peculiar*, Mark Twain's *Pudd'nhead Wilson*, and Charles W. Chesnutt's short story "Her Virginia Mammy" share a central plot in which Black women use red coral to transmit coded information about race that exposes the danger and limits of biological racism. And they do more than expose: these women also exist beyond that account of race by sharing and exchanging knowledge of coral that serves their own ends. The pages of these stories, then, fail to contain Black and mixed-race Black women within a US racial logic founded in slavery, a claim I test by turning to Stowe's *Uncle Tom's Cabin*, a text that most readers consider unlikely to carry and transmit any sources of Black Atlantic knowledge, given its notable reinforcement of the most violent US racial stereotypes, even as it agitates for abolition.[9] Might we, however, arrive at another reading of Stowe's novel by following the red coral that circulates through the hands of women—free and enslaved, white, Black, and mixed-race Black—at St. Clare's New Orleans household and that then carries us outward, into wider Atlantic currents of knowledge and experience?

FIG. 5.1. Diasporic identity: In the 1880 novel *The Grandissimes*, by US writer George Washington Cable, Palmyre Philosophe wears a red coral necklace, possibly to signal her complex identity as a New Orleans Creole woman of French and Senegalese descent. Albert Herter, "Silently regarding the intruder with a pair of eyes that sent an icy chill through him and fastened him where he stood, lay Palmyre Philosophe," 1899.

## PALMYRE PHILOSOPHE'S RED CORAL NECKLACE AND BLACK ATLANTIC RITUAL

Palmyre Philosophe's red coral necklace appears exactly once in George Washington Cable's *The Grandissimes*. In a scene significant and dramatic enough that American artist Albert Herter illustrated it for the 1899 edition of the text (fig. 5.1), the beautiful Palmyre, a New Orleans Creole woman of French and "Jaloff" (quite likely Senegalese) descent, is confined to bed by a gunshot wound and visited by Joseph Frowenfeld, a white immigrant to New Orleans and a stand-in for readers in his efforts to decipher the complexities of local Creole community and identity.[10] Upon arriving to dress Palmyre's wound, Frowenfeld finds himself transfixed. Palmyre, a "quadroon" who appears white to him, sits on her bed "silently regarding the intruder with a pair of eyes that sent an icy chill through him and fastened him where he stood."[11] She is dressed in a long white gown, "and a necklace of red coral heightened to its utmost her untamable beauty."[12]

This scene sharply stages the collision between two understandings of race that, as literary scholar Stephanie Foote observes, come into increasing tension as the narrative progresses, and in direct response to the impending US annexation of French Louisiana in 1803 (the period in which the novel is set).[13] On the one hand is a French colonial racial scheme, according to which one's race is multiply shaped by biology, culture, social position, and history; on the other is an emergent US account of race as strictly biological and thus "spectacularly ill equipped" for naming the complexly informed identities, such as Palmyre's, that emerged in colonial French New Orleans.[14] According to the former version of race, Palmyre is a free person of color who holds a clear, if somewhat marginal, social position in French Louisiana's Creole community.[15] According to the latter version of race—to which Frowenfeld subscribes and that is about to become the only version that legally matters under US annexation—Palmyre is Black, and thus stripped of citizenship altogether.[16] As US rule encroaches, Foote explains, Palmyre will no longer belong to a New Orleans Creole community comprising persons from a range of social ranks and racial backgrounds who share a common French cultural identity within French colonial Louisiana. Rather, "Creole" will come to mean biologically "pure" white, and Palmyre will thus be fully displaced, "deprived of even the marginal self-representation" she once possessed.[17] The only part of her identity that will matter is her Senegalese ancestry, which as scholar of African American literature Cassandra Jackson observes, is a result of the sexualized violence of white men against Black women.[18] With no prospect of incorporation into the US, Palmyre is "shipped away conveniently to France" at the novel's end.[19]

Within that context, the meeting between Palmyre and Frowenfeld described above positions Palmyre as a tragic "quadroon," the scene's poignancy deriving from her abject position within Frowenfeld's stringent US grid of racial intelligibility.[20] Palmyre is as Frowenfeld imagines her: a "monument of the shame of two races," a "final, unanswerable white man's accuser," who causes him to wonder "what ranks and companies would have to stand up in the Great Day with her and answer as accessory before the fact!"[21] We see Palmyre through Frowenfeld's combined attraction and compassion, which stem from an understanding of the increasing impossibility of her position. She "has been reconstructed as a type," Foote concludes, with no interiority or self-determination, as part of the novel's larger project of containing Creole "exoticism" on the way to allaying racial anxieties among Cable's late-nineteenth-century audience.[22]

There is no question that Cable's novel sustains the exclusionary violence of racial essentialism that rationalized slavery and other imperial projects. There is also room to interpret Palmyre in relation to meanings that far exceeded Cable's intentions or comprehension, and that position her as uncontainable within a distinctly US biological account of race. While Cable almost certainly knew nothing about the African and African diasporic practice of wearing red coral, the novel nonetheless cites it through the necklace Palmyre wears.

Women of African descent throughout the Black Atlantic frequently wore red coral to announce and celebrate sources of individual and social identity that alternately preceded or exceeded the commodifying logic of slavery.[23] Beginning at least as early as the late fifteenth century, red coral beads were valuable objects of social status and political power among Indigenous populations in coastal areas of present-day Senegal, Guinea, Liberia, Ghana, Benin,

Nigeria, and Angola.[24] In these places, Indigenous African value systems and oral traditions endowed the beads with such power that they became highly coveted items of dress and important currency in the transatlantic slave trade. Because European slave traders exchanged supplies of red coral for persons for over three hundred years, to Europeans red coral beads often constituted merely a "mercantile unit of human flesh" that, like other commodities, "determined the worth of slaves and provided the measure of their existence," Saidiya Hartman remarks.[25] But to the people forcibly taken from these and other places, the beads retained and acquired other meanings.

As historian of Africa and the Caribbean Steeve O. Buckridge reminds us in a study of the politics of dress among enslaved and free women of African descent in eighteenth- and nineteenth-century Jamaica, African-descended peoples did not forget the styles of dress and other cultural practices of their countries of origin.[26] Rather, they frequently enacted and adapted these practices in new contexts during and after slavery. Buckridge particularly observes that red "coral necklaces" were one of several items of dress through which enslaved women in Jamaica "created and maintained a vital link" to African roots, while "simultaneously . . . [resisting] the system of slavery that sought to rob them of their pride, their dignity and, most of all, their African identity."[27] In other words, the practice of wearing red coral was one of many sartorial ways that these women both created and "non-verbally" communicated and sustained complexly informed diasporic identities.[28]

That insight is supported and amplified by sources variously documenting or strongly suggesting that both enslaved and free women of African descent chose to wear red coral when participating in Jonkonnu, a Christmas holiday masquerade of African origins. As enacted publicly by Jamaica's free and enslaved Black populations, the parades and performances of Jonkonnu contested slavery by generating and nurturing complex social identities and relations.[29] Costumes were integral to that process, since each part was encoded with meaning. As Elizabeth Dillon explains, the dress of the so-called Jonkonnu "Set-Girls," groups of enslaved and free Black women, was particularly significant: their costumes "[announced] not only splendor and beauty but a form of social belonging" within different diasporic "kinship groups" of their own devising.[30] Each textile, color, and accessory helped these groups of women counter the "social death" of slavery by "[linking] one body in voluntaristic, flexible belonging to another."[31]

Red coral was apparently an important part of women's Jonkonnu costumes. When gathering at Christmas, and indeed on any other "extraordinary occasion," Black women in Jamaica "adorn their necks and wrists" with coral, observes Jamaican-born British slaveholder William Beckford in 1790.[32]

Writing about Jonkonnu on one of his Jamaican plantations during the early nineteenth century, British writer Matthew Gregory Lewis (also known as "Monk" Lewis) recorded that the female participants were notably "decked out with a profusion of beads and corals."[33] In 1838 merchant James Kelly observed that necklaces and bracelets of red "coral and cornelian" were especially prized components of women's Jonkonnu costumes.[34] And in the most widely known published color images of Jonkonnu in Jamaica during the nineteenth century, a set of hand-colored lithographs produced during the 1830s by Jewish Jamaican artist Isaac Mendes Belisario, many of the women wear necklaces, and sometimes earrings, that appear to be made of red coral.[35]

The practice was not confined to Christmas holidays or to Jamaica. In 1788 English visitor Peter Marsden observed a general fondness for coral necklaces among Black and mixed-race Black women who were "kept by the white men" as mistresses on Jamaican plantations.[36] At Saint Vincent and Trinidad "all ranks of the female slave population" wear coral necklaces, particularly on special occasions, reports Mrs. A. C. Carmichael in 1833 after five years in the West Indies.[37] And in the US South, a region that rightly belongs to a broader "circum-Caribbean" bound by shared circulations of goods, bodies, and knowledge, a number of disparate sources suggest the importance of coral to women of African descent. A Swiss traveler to eighteenth-century Virginia reported that "large numbers" of enslaved persons arriving annually from Guinea and British Jamaica disembarked "entirely naked," save for "corals of different colors around their neck and arms."[38] Eighteenth-century red coral beads have been excavated by archaeologists from plantation quarters where enslaved persons lived or worked, and from African burial grounds.[39] And in New Orleans red coral seems to have been an expression of status and Creole identity among Creole women of African descent. So indicates free Black woman and traveling hairdresser Eliza Potter (1820–after 1861) in her memoir when depicting a New Orleans "creole soirée"—possibly a quadroon ball—hosted by a wealthy Black Creole woman "who looked as though she might be white" and wore "white silk" and "a full set of coral" jewelry.[40] The attire strikingly resonates with that of Palmyre Philosophe and with the unidentified New Orleans Creole and free woman of color in a portrait produced during Cable's lifetime, Louis Nicholas Adolphe Rinck's *Woman in Tignon* (1844; fig. 5.2).[41]

The many women captured fleetingly by these sources did not share a unified "African" identity, or even a single African diasporic "group identity." As inhabitants of the Black Atlantic diaspora during the eighteenth and nineteenth centuries, however, they did express and pass on some shared "social memories," often through everyday practices that connected them across time and space, as Ogundiran and Saunders remind us. Many of these women could

FIG. 5.2. Creole identity: A red coral necklace is worn by this unidentified New Orleans Creole and free woman of color. Red coral ornaments may have expressed status and Creole identity among Creole women of African descent in nineteenth-century New Orleans. Louis Nicholas Adolphe Rinck, *Woman in Tignon*, 1844 (Hilliard Art Museum).

trace lineages or social connections to specific cultures in Africa where red coral beads held complex symbolic meanings and social valuations shaped by early encounters between Indigenous African value systems and European transatlantic commerce. And many apparently chose to wear red coral on occasions when clothing choices constituted self-definition and self-expression. While the full meaning of their practice must elude us, then, we must also notice its presence, persistence, and adaptation in variegated contexts in the Black Atlantic world across hundreds of years. That necessarily fragmentary genealogy of practice allows us to speculate that red coral ornaments carried and announced currents of a diasporic expressive culture beyond national borders.[42]

When we return to Palmyre's red coral necklace with this genealogy in mind, we might perceive her as something other than the tragic figure that Frowenfeld pities as a "monument of the shame of two races." Perhaps instead Palmyre silently exposes the limits of the entire system of biological race that would reduce her to Black or white, by signaling through coral that she bears and belongs to multiple histories inflected, but not defined, by US slavery. By this point in the novel Palmyre has already established herself as a skilled performer, and in one earlier scene she explicitly uses items of dress to announce her multiply informed identity. When meeting the captive "Jaloff" Bras-Coupé, Palmyre communicates to him that she is not white, as she appears to many observers in that time and place, but rather of shared African ancestry, by "[attiring] herself in a resplendence of scarlet and beads and feathers that could not fail the double purpose of connecting her with the children of Ethiopia and commanding the captive's instant admiration."[43] In this scene, beads help Palmyre tacitly but effectively affirm her identity. It seems just possible, then, that Palmyre's red coral necklace cites a long-standing practice through which Black and mixed-race Black women throughout the Atlantic world also generated and sustained identities that slavery could not contain.

Readers of other nineteenth-century US fictions of slavery and race have generally understood Black and mixed-race Black female characters in just the way that Palmyre has been understood, either as fully (and tragically) contained and explained by US racial ideology or as seeking to challenge it (unsuccessfully). But what if these women also just exist and communicate beyond that framework, through knowledge and experiences that derive from and announce other sources of Black identity?

In an unexamined plot, shared by the works I next examine, Black women mobilize red coral ornaments to contest a biological account of race. In its repetition across multiple texts, this plot becomes a repertoire, a script prompted by red coral because of its features and history, and that thus may carry and generate meanings beyond authorial understanding. That repertoire points toward the meanings that red coral acquired among different communities of Black women throughout the Black Atlantic diaspora. It encourages us to read fictional Black women more generatively, as characters who challenge race as biology *by* affirming their participation in other currents of knowing and living.

## US Literature's Red Coral Repertoire, ca. 1850–1900

In the convoluted plot of Epes Sargent's novel *Peculiar: A Tale of the Great Transition*, two pairs of red coral sleeve buttons become the most powerful material evidence that Clara Aylesford Berwick is legally white. Readers of the novel learn early on that Clara is the daughter of a white Northern couple killed in a steamboat explosion on the Mississippi River en route to New Orleans in 1848, when she was three years old. Yet within the narrative, Clara's racial identity comes into question in the chaotic aftermath of the explosion when she is captured, along with her "mulatto nurse" Hattie, by two slave traders who darken Clara's skin and sell her at a New Orleans slave auction.[44] There she is purchased by white supremacist Carberry Ratcliff. He deposits her at a New Orleans boarding school, where she is taken for a "quadroon" called Ellen Murray and cared for by Esha, the school's enslaved, mixed-race Black cook, for the next fourteen years, during which Clara loses her rightful inheritance to the villains who sold her. In 1863 Clara's purchaser reappears and, having failed to produce a white heir, presents the seventeen-year-old Clara with an ultimatum: marry him and thereby be recognized as white, as his wife and the mother of his child, or live out her days as a quadroon and his "white slave."[45]

By this point in the novel nearly everyone's fate depends on Clara's racial identity. For if Clara can prove her legal whiteness without having to marry Ratcliff, she will gain her freedom, regain her inheritance, and have the right to marry her chosen suitor. The villains who sold and bought Clara will receive their long overdue punishment for falsifying her racial identity, and both of Clara's suitors will triumph—one by marrying Clara, the other by avenging the death of his first wife (whom, it turns out, Ratcliff had also illegally enslaved). *Peculiar* is a novel about evidence, about what counts as proof, why, and for whom, as literary scholar Justine S. Murison insightfully argues, and the plot involving Clara is about racial evidence in particular. The question is how to establish Clara's lineage as "pure Saxon" when her body alone is not reliable evidence, and to that end Clara and her allies summon "a collection of testimony," both spiritual and material, in her defense.[46]

Prominent among the material evidence of Clara's racial identity, though Murison does not discuss it, are the two pairs of red coral sleeve buttons that Clara was wearing during the steamboat accident, and that disappear in the aftermath, only to reappear nearly two decades later and circulate in the most improbable ways and places.[47] One pair of the buttons first reappears at the boarding school when Clara, age seventeen and living variously as "Ellen" or "Perdita" ("the lost one"), first learns that Ratcliff owns her: while she exclaims that she now knows her "true place in the world," and that "the mystery is solved"—that she is "a slave"—Esha produces a pair of coral buttons, which silently reminds readers of Clara's actual parentage, and thus that nothing is "solved."[48] Clara then wears one button when, continuing to question her identity, she decides "to vindicate her own right to freedom by showing she was not born to be a slave," partly by dressing "to dazzle and to strike," which involves pinning the button to her collar.[49] The strategy succeeds when the button slips from her clothes and into the possession of her suitors, who identify it as belonging to a set engraved with the initials "CAB" for Clara Aylesford Berwick.[50] In their final appearance the buttons enter the legal record as official proof of Clara's whiteness: in a formal deposition the boarding-school owner identifies them as the same ones worn by the child enrolled by Ratcliff, thereby releasing Clara from bondage by proving that she was born to white parents.[51]

On their wild circuit through the narrative, the buttons come to serve characters as incontestable proof of Clara's "pure" white blood, while they increasingly signal to readers the dangers and irresolvable quandaries produced by searching for biological proof of race.[52] The plot leans so much on the accurate identification of these four tiny coral buttons—which somehow survive a steamboat explosion to circulate through the hands of multiple characters in different locations across a period of fourteen years—that the more seriously the characters take this coral evidence, the more implausible it becomes to readers.

In this way the novel marshals a popular nineteenth-century meaning of coral as a signifier of the drive to taxonomize and of the irresolvable quandaries produced by that project. That meaning of coral emerges partly from the category crisis surrounding coral within European and American natural history, which I explored in chapter 3. And it renders red coral an apt material manifestation of the narrative's implicit challenge to the politically powerful "science" of race-as-biology.[53] And that might be all there is to say of how red coral adornments signify in *Peculiar*, except for the important fact that the complex double meaning of coral as evidence both of whiteness (within the text) and of the limits of US racial classifications (to readers) depends entirely on how the coral is strategically deployed by mixed-race Black women.[54]

In *Peculiar* mixed-race Black women use red coral to exercise what Fielder calls their "racializing potential," their ability "to place" people "into . . . racialized relations" within a culture that defines race as biological—and in so doing to signal the limits and dangers of that definition.[55] There are two people in *Peculiar* who render red coral racial evidence: Clara's caregiver at the boarding school, the enslaved Esha, whom the narrator describes as "somewhat Caucasian" in appearance, who speaks Arabic and who was born "Ayesha" in Africa to a Muslim father; and Clara's childhood nurse, the "mulatto" Hattie, who turns out to be Esha's "long-lost daughter."[56] It is Esha who, possibly suspecting Clara's whiteness, first identifies one pair of the buttons as belonging to Clara and gives the pair to Clara at the moment when Clara declares "I'm a slave," thereby making it available for Clara to wear in the presence of her suitors.[57] Next it is Hattie who persuades Clara's suitors to examine the coral more closely and perceive that it belongs to a set once worn by the child known to one of them as incontestably white; coral thus becomes the missing evidence of Clara's racial whiteness.[58] Finally, it is Esha's testimony that ultimately frees Clara from slavery by corroborating Mrs. Gentry's legal deposition that the buttons have always belonged to Clara.[59] At each moment, then, the ability of white characters to interpret the buttons as evidence of Clara's "true" race and place depends on the actions and words of Esha and Hattie and thereby so too does the reader's perception of the buttons as a silent indictment of this culture's pride and certainty in discerning what one of Clara's suitors calls "slight differences of color."[60]

That repertoire repeats with some variation across two later and more widely read nineteenth-century US fictions wherein it likewise upholds what Eric J. Sundquist calls a narrative strategy that "[exposes] to view while indulging" a US racial hierarchy based on biology and sponsoring the most violent forms of "power and mastery."[61] Once again, coral operates this way only because of how mixed-race Black women deploy it.

In the considerably more popular novel *Pudd'nhead Wilson* (1893), by Mark Twain, for instance, a red coral necklace becomes crucial material evidence that Valet de Chambre (called Chambers) is white, while increasingly signaling to readers the perils, contingencies, and impossibilities of searching for biological whiteness. Chambers is born on the same day in 1830 in the same household as Tom Driscoll, son of Percy Driscoll, a slaveholder living along the Mississippi River south of Saint Louis.[62] Readers realize early on that Chambers is in fact the biological son of Roxana, or "Roxy," the Driscoll's enslaved "mammy," who along with Chambers, appears white yet is Black according to the legal and cultural codes of this time and place.[63] Yet the lives and welfare of nearly all the characters increasingly turn on the uncertainty surrounding the racial identity of both Chambers and Tom. For when the two are seven months old Roxy swaps them by exchanging their articles of dress, including Tom's signature coral necklace, by which Tom's father knows him from Chambers.[64] The exchange enables Roxy to perform as "mammy" to her own son and mother to Tom, thereby preventing her biological son from being "sold down the river" for over twenty years.[65] Chambers's ability to live for more than two decades as a descendent of "the best blood of the Old Dominion," along with readers' understanding that the novel deeply questions this culture's obsession with blood "purity," depend on Roxana's red coral repertoire.[66]

A third iteration of that repertoire appears in Charles W. Chesnutt's short story "Her Virginia Mammy" (1899): a mixed-race Black woman positions a red coral necklace as incontestable proof of Clara Hohlfelder's pure "white" blood, even as readers realize that Clara is legally Black.[67] The story begins with Clara's refusal of her white suitor's marriage proposal on the grounds of her uncertainty about her biological lineage, despite his assurances that he would marry her even if she were Black.[68] Upon her adoptive mother's death, however, Clara chances on a coral necklace among other "trinkets" that she scrutinizes for clues to her identity, though these objects "tell [her] nothing."[69] It requires the chance appearance of one Mrs. Harper, a woman of "clear olive complexion" who turns out to be the only living person who can read these materials as racial evidence.[70] Clara presents the items to Mrs. Harper, disclosing the coral necklace with the words, "perhaps your keen eyes can find something in that." The story continues, "The older woman gave but a glance—a glance that added to her emotion": the necklace confirms beyond doubt, she explains, that Clara is "the best blood of the Old Dominion," the daughter of a white couple for whom Mrs. Harper served as "mammy."[71] The news enables Clara to marry—even though readers realize, along with all of the main characters except, it seems, for Clara, that Mrs. Harper is Clara's biological mother.[72]

A red coral necklace thus powerfully sustains a performance that most of the characters recognize as such, even while it propels the story's plot by effacing Mrs. Harper's maternity and enabling Clara to marry. That shared awareness of the performative nature of Mrs. Harper's racial repertoire distinguishes Chesnutt's iteration from Twain's and Sargent's, and arguably renders it a more powerful critique of racial biology. Readers of this tale cannot escape Chesnutt's indictment of a culture organized around such an obviously fictive story that race fully resides in individual bodies, where it can be found and assessed if only one looks hard enough.[73]

But when we *also* read Chesnutt's tale within the context of other narratives that feature mixed-race Black women deploying red coral to expose the limits of this story of race, then we have to ask whether all these texts are doing something other than critique and exposé. They all depict a material practice that strikingly resonates with one carried out by Black women in African and African diasporic communities who, like Mrs. Harper, "find something in" red coral ornaments. Even in the case of the white writers who feature in this chapter—those who were less likely to know or promote African diasporic cultural knowledge—these stories nonetheless evoke and transmit remnants of a material practice of coral carried out by women of African descent who lived in different places and circumstances yet shared historic and ongoing entanglements in Atlantic circulations of goods, bodies, and ideas.[74]

For if we approach red coral as a "scriptive thing," then we must notice how Mrs. Harper's interpretation and use of red coral resonates deeply with that of Esha, Hattie, and Roxy. That interpretation, in turn, points us toward a cultural understanding of coral within communities of Black and mixed-race Black women for whom red coral ornaments announced identities beyond the continent and its racial ideologies.

## STOWE, CORAL, AND RACE

At the first appearance of red coral ornaments at Augustine St. Clare's New Orleans household in *Uncle Tom's Cabin*, the gestures and glances of Black and mixed-race Black women suggest that red coral holds social meaning to them. The enslaved "quadroon" woman, Jane, uses a pair of red coral earrings to communicate her position among "the higher circle" of enslaved persons at St. Clare's.[75] When speaking with Prue, a Black enslaved woman from a neighboring estate, Jane begins "dangling, as she spoke, a pair of coral ear-drops," a move that elicits Prue's retort: "Ye think ye're mighty fine with them ar [earrings], a frolickin' and a tossin' your head, and a lookin' down on every-body."[76] Later Jane's earrings communicate her difference from her would-be suitor, an enslaved Black man called Adolph. When Adolph declares Jane's earrings "bewitching" and asks whether she plans to wear them at the following night's ball, Jane understands that he is asking to dance with her and rebukes his "impudence," both verbally and by "tossing her pretty head 'til the ear-drops twinkled."[77]

*Uncle Tom's Cabin* is an "archive of repertoires," observes Bernstein, for how much of the text happens wordlessly in glances, gestures, behaviors, and other nonverbal performances, many of them scripted by racialized materials.[78] The novel's vast catalog of such materials includes Eva's locks of hair, Tom's Bible and coin, and Mrs. Shelby's handkerchief, objects that have received signifi-cant scholarly attention, mostly for how they silently reinforce the ideology of biological race that Stowe herself endorsed during the 1850s.[79] Yet might several overlooked, early, and fleeting scenes of red coral, a material distinct from these other objects in its oceanic origins, nature, and history, as silently carry a different script into the text?

The scenes in which Jane uses coral to announce the complexity of her identity may be setting the stage for the chaotic arrival of "blackest" Topsy, who soon after comes into the St. Clare household as Augustine's "present" to his New England cousin Ophelia, apparently bringing her own conception of red coral.[80] On the night Topsy arrives, "dreadfully dirty, and half naked," Jane greets her with disgust, which Topsy effectively scorns by merely "scanning, with a keen and furtive glance of her flickering eyes, the ornaments which Jane wore in her ears."[81] The following morning Topsy declares to Ophelia that she has stolen a different pair of red coral earrings, those belonging to the enslaved quadroon woman Rosa, along with the signature red coral necklace worn by Augustine's daughter, the angelic Eva St. Clare.[82] What is more, Topsy confesses, she has burned these coral ornaments to destruction, and for nothing other than pure wickedness—which turns out to be untrue when, moments later, Rosa and Eva appear wearing the jewelry in question, untaken and intact.[83]

Topsy's false confession upends Ophelia's vaunted "experiences and opinions," and particularly the New England "prejudice of caste and color" that Stowe describes as Ophelia's characteristic trait in *A Key to Uncle Tom's Cabin* (1853).[84] For Ophelia finds herself "perfectly bewildered" at Topsy.[85] She declares Topsy a "puzzler" who eludes "her rules for bringing up"; in need of time to "[arrange] her ideas further on the subject," Ophelia shuts Topsy in a closet.[86] But the ensuing conversation, in which Ophelia seeks counsel from Augustine, indicates that Topsy has forever disarrayed Ophelia's thinking about Black identity.

Stowe's conviction that Black persons are reducible to racial "types" is everywhere on display in this novel, and particularly so, scholars have suggested, in conversations between Ophelia and Augustine on the nature of Topsy. Augustine first introduces Topsy to Ophelia as a "fresh-caught specimen."[87] He maintains that categorical thinking to the end, describing Topsy as a "very common" type.[88]

There is much evidence that Augustine's views are Stowe's during the period when she wrote the novel. Yet after the coral necklace episode, Ophelia can no longer maintain them. Of Topsy she declares, "I never saw such a child as this" and "I don't know" what to do with her.[89] These words suggest that, to Ophelia, Topsy is now neither "present" nor "specimen" but "puzzler," living evidence of the limits of Ophelia's own grid of racial intelligibility.[90] Instead of the threatened whipping, Topsy receives an education from Ophelia. Ophelia then purchases Topsy from Augustine to protect her from being "hustled off to auction" in the event of his death.[91] And when he does die, Ophelia brings Topsy to New England and supports her decision to leave the country for good. Baptized "by her own request," Topsy exits the narrative and the nation, taking up missionary work in Africa.[92]

Because Topsy's "freedom involves her submission and removal," the conclusion to her narrative shows "the limits of Stowe's imagination," observes literary scholar Samuel Otter in an insightful assessment of "Stowe and Race."[93] Moreover, literary critic and scholar of Black Feminism Hortense J. Spillers powerfully interprets Topsy and other Black women in the novel as voiceless, mere "ciphers" "for all their sporadic 'talking.'"[94] That account of Black women in nineteenth-century white writing has been widely applied by other scholars, including to Cable's Palmyre.[95]

But if meaning also resides beyond language and in gestures and glances provoked by particular materials because of their histories, then perhaps Ophelia's encounter with Topsy and coral deserves further consideration in our assessment of Stowe and race. For while Topsy's life is unquestionably constrained by the racist logic that refuses Black self-determination, that logic is also momentarily pierced, its limited contours made visible, if only fleetingly, by the understanding that Topsy exceeds it.

In this way Ophelia's encounter with Topsy can be read alongside Frowenfeld's with Palmyre: in both cases, Black women mobilize red coral to affirm identity beyond the frame. Because Ophelia's relationship with Topsy properly begins with Ophelia's admission, "I don't know," Stowe's narrative may even fleetingly imply that white epistemic humility must be the foundation of more just relations across racial boundaries.[96] If that is the case, then the coral of *Uncle Tom's Cabin* also gives us one way out of a system of knowledge according to which race is a fixed and biological fact.

INTERCHAPTER

# Topsy in Coral

The strange journey of Eva's coral necklace through the visual culture of *Uncle Tom's Cabin* silently affirms the power of Topsy's coral repertoire. We know that, within the novel, Eva never removes her coral necklace. In fact, in answer to Ophelia's surprise when Eva enters wearing the necklace that Topsy has allegedly stolen and burned, Eva exclaims, "I've had it on all day" and "all night."[1] Closely following the words on the page, many artists who produced popular images of Eva and Topsy touching, or even embracing, chose to depict Eva wearing a coral necklace (Corbaux 1852; fig. 5.3).

But in some images of the two girls, the red coral necklace finds its way to Topsy (anonymous 1880; fig. 5.4 and Bingham 1857; fig 5.5). Why did some nineteenth-century artists choose to portray Topsy, rather than Eva, in coral?

One likely answer is that these artists were influenced by the long-standing convention, discussed in this book's first chapter, of allegorizing "Africa" as a black woman in coral (fig. 1.8). Certain scenes in the novel could plausibly have coaxed artists toward adopting this convention when representing Topsy. For example, when Topsy and Eva first meet—just after the discovery that Topsy has not stolen Eva's necklace—they stand facing each other, "the Saxon, born of ages of cultivation, command, education, physical and moral eminence; the Afric, born of ages of oppression, submission, ignorance, toil and vice!"[2] Topsy's status as "the Afric" seems especially prominent in J. A. Bingham's 1857 portrait (fig. 5.5), which follows Ripa's seventeenth-century iconographic instruction quite exactly by giving Topsy "a Necklace of Coral; and Pendents of the same, at her Ears."[3]

EVA AND TOPSY.

I love you because you haven't had any Father, or Mother, or Friends, —
because you've been a poor abused child!

*Vide "Uncle Toms Cabin"*

FIG. 5.3. "I've had it on all day" and "all night": In Harriet Beecher Stowe's *Uncle Tom's Cabin*, the angelic Eva St. Clare wears a signature white dress and red coral necklace, attire noticed and portrayed by artists who depicted particularly popular scenes in the novel, such as the one in which Eva touches Topsy and converts her to Christianity. Louisa Corbaux, "Eva and Topsy," 1852 (Library of Congress, Prints and Photographs Division).

Likewise, Bingham's portrait resonates strikingly with porcelain figurines featuring "Africa" embraced by "Europe" (Meyer 1760; fig. 1.6): in portrait and porcelain alike, a white child holds a black child with coral. Considered in this way, images of Topsy in coral confirm her status as a "specimen" or "type," thereby reinforcing the racial logic nearly everywhere on display in *Uncle Tom's Cabin*.

Yet might the artistic choice to portray Topsy in coral instead suggest the degree to which she threatens to disrupt that logic? It seems at least possible that this choice, so at odds with the novel's descriptions, constitutes an act of reinscription, an effort to restore Topsy after the fact to a fixed place within a system of racial classifications and convictions, and thus evidence that she destabilizes that system in the narrative. When first asked her age, place of birth, and parentage, Topsy replies that she "never was born" but just "grow'd."[4] The

LITTLE EVA CONVERTING TOPSY.

"In that moment a ray of real belief, a ray of heavenly love, had penetrated the darkness of her heathen soul."

FIG. 5.4. Topsy in coral: In some images that artists produced of the famous "conversion" scene from *Uncle Tom's Cabin*, the red coral necklace appears on Topsy. In the novel itself, however, the necklace belongs to Eva, and Topsy never so much as touches it. "Little Eva Converting Topsy" (1880–90?). Based on 1852 engraving (Library of Congress, Prints and Photographs Division).

FIG. 5.5. Topsy wears a full set: Notice the coral jewelry at Topsy's ears, neck, and wrist in this illustration of a scene from Harriet Beecher Stowe's *Uncle Tom's Cabin*. Illustrators of the novel may well have been influenced by the allegory of "Africa" as a woman wearing coral: compare this portrait to Eckhout's *African Woman* (shown in Fig. 1.5.). J. A. Bingham, "Topsy and Eva," 1857 (Harriet Beecher Stowe Center, Hartford, Connecticut).

implication is that Topsy both self-propagates and grows unpredictably and without design, for this particular line in the novel quickly gave rise to the British expression, "to grow like Topsy," which has been used since the mid-nineteenth century to designate "something that seems to have grown of itself without anyone's intention or direction."[5] Could Topsy, too, "come from coral reefs" like the speaker of Nichols's "Web of Kin" (1985)—hybrid, transitory, and always irreducible to slavery's logic?[6]

Nonteleological growth has, of course, long been an imagined quality of coral. Perhaps this quality also informs Topsy's silent affinity for coral in the text. For all Topsy's silence, then, figures of Topsy in coral may indicate that some readers heard a provocative script of self-determination in her own account of her origins and nature.

# Coral Temporalities

The origins and nature of coral are no longer a familiar topic of wide-ranging reflection and debate in US classrooms and living rooms, poetry and song, news reports, novels, and sermons. And when reefs come to mind these days, they carry different associations—warming seas, species loss, planetary crisis. It was not always this way. Until not so long ago, reefs raised and reshaped long-standing political ideas about identity, community, labor, race, belonging, and even time itself.

In the final chapter of *The Edge of the Sea* (1955), Rachel Carson reflects that "the flow of time" is always "obliterating yet containing all that has gone before."[1] Time, according to this construction, is multilayered. The present is never separable from the past but always constituted by it, even while moving toward the future. The natural object that embodies these elusive characteristics of time, for Carson, is coral.

Standing on the coral ground of the Florida Keys, Carson becomes "strongly aware of the old, dead reef beneath my feet, and of the corals" who built it. "All the builders now are dead—they have been dead for thousands of years—but that which they created remains, a part of the living present."[2] A reef grows by both obliterating and containing the builders: the polyps necessarily die, but their labors and lives are not erased because their limestone exoskeletons are quite literally the material of the reef. The "living present" of coral, then, is largely the "remains" of its many pasts.

As such, the reef reorients Carson to her own present. Coral makes Carson newly and "strongly aware" that the present carries the past. It attunes her, in other words, to her place within a multilayered time that passes and accumulates at once.[3]

Carson's coral temporality involves a relation to the past that Dominick LaCapra might call "humanistic." In a defense of the humanities at a moment when teaching and research in fields such as literature and history are increasingly imperiled by public insistence on education that prioritizes immediate, quantifiable utility, LaCapra points out that the "use" of the humanities has never been its "deliverables" or "technological payoffs."[4] One of the more important values of a humanities education is the relation to the past that it alone cultivates: "the humanistic relation to the past assumes the past is not simply past but in significant ways still part of the present with implications for the future," LaCapra reflects.[5] By assuming this relation we acknowledge that our present is never free of the weight of a continually unfolding past in which we necessarily participate.

The concept of a layered present is not new, and it is adaptable to many purposes.[6] Yet without it, LaCapra observes, we are unlikely to successfully analyze or address complex problems that repeat with variation over time, such as trauma, oppression, justice, community, and responsibility.[7] Any hope for a more just present and future depends on prioritizing that relation, according to many scholars of the entwined histories and legacies of colonialism, slavery, and capitalism. We might think, for example, of anthropologist and historian Michel-Rolph Trouillot's observation that "there is no perfect closure to any event"; of the "afterlives" of slavery that scholar of African American literature and culture Saidiya Hartman identifies as "skewed life chances, limited access to health and education, premature death, incarceration, and impoverishment"; or of the insights of historians of slavery and capitalism Destin Jenkins and Justin Leroy that "the temporality of racial capitalism is one of ongoingness, even if its precise nature is dynamic and changing" and that "it is a process, not a moment."[8]

These and other scholars in many disciplines and fields call for wider recognition that we are implicated in "the past, its processes, and its artifacts"— that, as literary scholar Jennifer Fleissner explains, the past "continues to make claims upon us."[9] Yet we might never achieve that recognition without first feeling what a reef prompted Carson to feel, that the present contains the past. For it is one thing to theorize this relation and quite another to become, as Carson might say, "strongly aware of" it—to know it tangibly, so that it might direct our everyday thoughts and actions.

If we seek a metaphor for the presence of the past, we can scarcely do better than coral, as it was perceived and experienced by Carson and many earlier Americans who feature in this book. For them, coral gave much-needed shape and form to the past in the present. I return briefly now to three of these writers, each active during a different decade of the nineteenth century, to reframe their reflections on coral and labor as reflections on temporality as well, and specifically on the recognition of responsibility to others that might emerge from understanding ourselves within multilayered time.

During the 1830s, Sarah Josepha Hale found that a coral branch specimen disrupted the everyday course of life by reminding her not only of geological time, the scale on which coral is produced by countless generations of laboring polyps, but also of her own position within the continuum of past, present, and future. From now on, "when coral I see," states the speaker of Hale's "The Coral Branch" (1834), I pause to remember "that, dying, I should leave / Some good work here / My friends to cheer / When o'er my tomb they grieve."[10] By imagining her life and labor as a reef, the speaker understands herself and her labor within a present that contains the past; this is why her present "work" will live beyond her to shape the present of others yet to be.

Later in the nineteenth century, writing in 1856, historian and translator Elizabeth Wormeley Latimer located tremendous political potential in the layered temporality of living coral. Drawing on a Darwinian understanding of coral's growth, Latimer envisions human history as a reef in which each individual polyp is simultaneously "growing out of" a previous generation and "elaborating"—both designing and making—the "conditions" for a future one.[11] Within that model of a reef each individual is connected to a vast variety of others, present and past, through their labor; so many, in fact, that the movement of one polyp could alter the course of history. Coral thus teaches humans, according to Latimer, that we are responsible for countless others beyond ourselves, and are obliged to continually nourish the numerous "relations" in which we stand to them, across and between generations, particularly if we wish to create a more just future.

And on the first day of the twentieth century, African American minister Abraham Lincoln DeMond drew on the figure of a coral reef to describe the layered temporality that he considered the necessary condition of racial justice in the US. DeMond's New Year's Day oration at Dexter Avenue Baptist Church urges Black listeners to participate more actively in the public and political life of the country, to become "identified with all the interests of the nation, weaving our destiny with hers."[12] Yet he knew that his vision for the nation's future could not come to fruition in the absence of popular acknowledgement that Black people *are already* woven into the nation's history; indeed, that the nation's present is fully "enabled" by nearly three hundred years of Black contributions, which have been excised from the popular historical narrative in the service of upholding a myth of national origins in the "brilliant deeds" of the Puritans.[13] While these deeds have been "enshrined" in "poetry and prose, in music and song and art," the "Negro's story" has yet to be told, declares DeMond.[14]

To help listeners grasp and act on the constitutive role of Black people in US history, DeMond evokes a coral reef. Black life, he explains, has "entered into the structure of this fair Republic of the Western World. 'The coral insect but an atom gave / To help uprear the pile he ne'er could see; / But now it towers above the topmost wave, / He has a part in mansions yet to be." Black people have already "entered into the structure" of the Republic, DeMond affirms. Generations "gave" of themselves without even being able to see what they were building. Broader acknowledgment that the reef/Republic arose from their work is the condition of Black participation in the country's future.

DeMond's coral metaphor positions Black listeners within a national present that contains, rather than elides, the past, which is the only position from which, he imagines, they would be willing to "weave" themselves into the nation's future. Coral, in other words, helps make tangible the present's indebtedness to the past, and particularly to past labors, which is why a reef metaphor also inaugurates DeMond's brief history of the US as a country that "slave labor enabled," from "Jamestown in the year 1619" when "the Negro element in American life made its first appearance," "his strong iron arm of labor" preserving the colony from failure.[15] While DeMond goes on to name many Black contributions beyond physical labor, his recasting of Jamestown ought to help us see why 1619 still matters: in DeMond's time, as in our own, a narrative of US history that centers white "deeds" perpetuates the violent reduction and erasure of Black life and labor that rationalized slavery.[16] Within DeMond's coral temporality, by contrast, 1619 is present in 1900, there to be reckoned with for any real "weaving" to happen.[17]

From our present, twenty-first-century vantage point, it could seem odd or unlikely that coral, of all materials, once manifested the past, and particularly human pasts, so forcefully and for so many purposes. Today coral itself vanishes due to human-driven climate change and reefs too easily signify all that human industry endangers, including human life, which ultimately depends on coral reefs.[18] Relatedly, our culture has historically mistaken the sea in its vastness as a region remote from human affairs, or even immune to human intervention: we might think of Melville's "trackless" ocean, which resists or dissolves any human effort even to mark or to map.[19] As American studies scholar Jamie L. Jones has pointed out, drawing on the work of Hester Blum, oceanic metaphors can tend to suggest the sea's timelessness and abundance, thereby furthering industrial extraction and contamination on the false premise that the sea will always remain plentiful and unchanging.[20]

Yet as Jones also observes, a number of scholars working in Black studies have recently begun to suggest that the sea can also abet anticolonial thinking and action.[21] In the work of Christina Sharpe, for example, the figure of "the wake," defined as a "region of disturbed flow," offers a vital model for attuning ourselves to slavery's enduring presence: it evokes both the forced Atlantic crossings and drownings of the slave trade *and* the persistence of lives that slavery shaped, yet could not contain in its hold.[22] Sharpe's work suggests that untold histories of the sea and sea materials may yet offer us new and much-needed models and metaphors for assuming a more accurate and responsible relation to others past and present, which is finally what coral offered Hale, Latimer, DeMond, Carson, and many other Americans whose words and works this book explores.

And so this coda is also an argument for the indispensability of a historically oriented environmental humanities. As the field of environmental humanities has emerged, at least in the US, it has shifted toward the study of contemporary texts alongside contemporary problems. This shift makes sense: our current social and environmental crises are so pressing that a turn to the past can seem like a turn away from what matters.[23] But a careful reader of nineteenth-century American writing in the age of climate crisis will also, as Michelle C. Neely observes, find that it offers alternative models of living in relation to others that, if widely adopted now, might finally produce a more truly just collective.[24]

A willingness to approach the past in this way is, after all, a rejection of the triumphalist arc of history, Neely points out, and it resists what Fleissner might call a misguided "certainty that the past exists in order to irradiate the virtues of the present."[25] It is also, relatedly, a recognition that, as Elisa Tamarkin suggests, a past text need not always appear immediately useful in order to be profoundly relevant to us.[26] In another century's preoccupations with coral we might unexpectedly find what we most need to make sense of and navigate our own moment.

In the end, one of the most enduring values of coral for humans may be that it tends to slow us down. Living coral remains profoundly arresting whenever we are fortunate enough to encounter it. Shockingly vibrant, it stops us so that we wonder and think. Perhaps we might also allow it to reorient us in our relationship to the past and the living present as coral has done for so many others before us.

# Acknowledgments

It is a privilege to write a book under any circumstances; I am all the more aware of that privilege for having written much of this one during a pandemic. The unforgivable decisions that made COVID-19 what it was and is in this country have stamped this book indelibly. In light of all that has been delayed, damaged, worn away, or lost, it seems only right to begin these acknowledgments by acknowledging my good fortune to have written it at all.

I am profoundly thankful to the English Department at Miami University, where I have been pleased to work since 2014. I am especially grateful to our brilliant department chair, Madelyn Detloff, for steering us through unimaginably trying times with unstinting goodness, and to the staff who make our working lives possible, particularly Amy Toland, Sarah Broome, Rachel Treadway, and Denise Roell. For above-and-beyond mentorship, I am grateful to Mary Jean Corbett, Andrew Hebard, Tim Melley, Liz Wardle, and Patrick Murphy, Literature Program Codirector of the Year, this and every year. For friendship and support over many years I am thankful to all my colleagues, and especially Sara Austin, Jim Bromley, cris cheek, Yu-Fang Cho, Erin Edwards, Daisy Hernández, Cheryl Johnson, Katie Johnson, Cindy Klestinec, Tim Lockridge, Margaret Luongo, Anita Mannur, TaraShea Nesbit, Kaara Peterson, and Cathy Wagner. Beyond the department, I am grateful to Associate Dean Renee Baernstein for granting me time away from campus for additional research and writing, and to Miami's Office of Research and Innovation for grants supporting the manuscript's preparation for publication. I am also grateful to the staff of Miami Libraries, and especially Mark Dahlquist for his support of my teaching and research, for sharing my fascination with the Coral Builders Society, and for finding invaluable information on their founder, Frank Lee. Thanks, too, to the many undergraduate and graduate students who inspire me.

Several institutions and foundations supported this book through fellowships and grants that enabled me to take time away from teaching and service obligations at Miami in order to conduct research and write. A year-long fellowship from the National Endowment for the Humanities during 2017–18 was instrumental to this project. During that year I dramatically expanded my research at several US libraries, archives, and museums, where I met and exchanged work with other scholars in many fields and disciplines that this book engages and drew on the expertise of countless research librarians and curators. I spent most of that year as NEH Fellow-in-Residence at the Huntington Library in Pasadena, California, for which I owe particular thanks to Director of Research Steve Hindle and to a vibrant group of Huntington fellows, many of whom offered influential comments on parts of the manuscript, particularly Janet Browne, Mary Mendoza, and Catherine Roach.

During 2017–18 research for this book was also supported by a Mellon Foundation short-term fellowship from the Library Company of Philadelphia and Historical Society of Pennsylvania; a Jack Miller Center short-term fellowship from the Newberry Library; and a Charles H. Watts Memorial short-term fellowship from the John Carter Brown Library. Each opportunity connected me with different groups of people who advanced my understanding of coral's literary and cultural history. I am grateful to staff at each of these institutions, and I owe particular thanks to Library Company librarian Jim Green and curator Connie King, whose vast and detailed knowledge of the collections is fearsome to behold.

These grants, along with others that I had earlier received for my first book, *Liquid Landscape*, allowed me to visit multiple additional institutions whose staff expertly located various coral objects, images, and texts. At the Library of Congress in 2014 I first encountered Sarah Josepha Hale's remarkable song about coral, thanks to Tom Mann and other reference librarians. At the Smithsonian National Museum of Natural History coral curator Steve Cairns opened drawer after drawer of coral specimens retrieved by the US Exploring Expedition. At the Museum of Fine Arts-Boston curator Kathryn Gunsch gave me a tour of the Benin bronzes, and curator Caroline Cole brought out some of Oliver Wendell Holmes's coral stereo cards, as well as a stereoscope to view them. While at the Harriet Beecher Stowe Center I caught sight of a portrait of Topsy in coral, and I am grateful to Director of Collections Beth Burgess for taking endless photos of it. At the American Antiquarian Society curators Laura E. Wasowicz and Elizabeth Watts Pope supplied me with so many rare texts about coral in a single day that I wondered if I was dreaming. And on a visit to the Winterthur Museum in 2018 one gracious curator allowed me to hold and shake a "coral and bells" (gently and with gloves on).

The latter stages of writing this book were greatly supported by a fellowship during fall 2021 at the Rachel Carson Center (RCC) for Environment and Society in Munich, Germany, where I revised the manuscript in response to readers' reports and new insights from an international group of RCC scholars in multiple disciplines. My particular thanks are due to Director Christof Mauch and Managing Director Arielle Helmick, and to the excellent RCC Staff, including Lena Engel, Maximilian Heumann, Sophia Hörl, and Laura Mann. I am grateful to those Fellows and graduate students who attended my work-in-progress presentation and offered valuable suggestions, including Regina Bichler, Ron E. Doel, Carmel Finley, Sasha Gora, Davide Orsini, Claudia Pessina, Miles Powell, Michael Schüring, Sevgi Mutlu Sirakova, Nike Stolpe Wikström, and Nina Wormbs, and to others who enriched my time at the center, including Rae Choi, Sarah Daw, Rebecca Kariuki, Francesca Mezzanzana, and Jonatan Palmblad. The semester that Gene and I spent in Munich was restorative and transformational; *vielen herzlichen Dank an alle.*

My work on *Coral Lives* brought me back to the Huntington Library in summer of 2022 to finish the book where it first began in 2013—in conversations with other Huntington Fellows, and particularly Bill Brown—before taking shape in many summers since. For supporting all these years at the Huntington, formal and informal, I am grateful again to Steve Hindle and to the Huntington Library staff, particularly Leslie Jobsky for finding the books and Juan Gomez for finding the space. One of the many great gifts of the Huntington Library is the community of people who live in, or regularly return to, Los Angeles because of it. Sarah Hanley and Malcolm Rohrbough have been the kindest friends and most generous hosts. Pat Smith and the late and incomparable Bob Smith have been the most wonderful neighbors. Because of these and many other people at many other times, I now know that the best ideas arrive on walks to see the Caltech turtles, or in the summer silence of Mahogany Row.

I owe special thanks to a number of scholars, many of whom I am fortunate enough to call friends, who have shaped this book as it developed, in some cases by taking time to read and comment on part or all of it. Nearly every word has been read (and reread) by the members of the Early Americanist Writing Group: Angie Calcaterra, Travis Foster, Greta LaFleur, Wendy Roberts, Kacy Tillman, Abram Van Engen, and Caroline Wigginton are still my most lively, demanding, and generous audience, and they serve as good evidence that the way forward is by listening to others (unless it's Carwin). Beyond that group, I am grateful to the sage suggestions and insights of Hester Blum, Andrew Hebard, Dana Luciano, Dominic Mastroianni, Janet Neary, Michelle Neely, Scotti Parrish, Liz Polcha, Matt Salway, Elizabeth Stockton, Dan Vandersommers, and Audrey Wasser. Cristobal Silva continues to remind me of what matters most in scholarship and outside of it, too. Elisa Tamarkin has steadfastly supported and influenced my work in countless ways for almost twenty years. For conversations that shaped my thinking, or in some way cheered me on, I am grateful to many others across the profession, including Carla Bittel, Alejandra Dubcovsky, Bert Emerson, Sam Ewing, Adrian Finucane, Lauren LaFauci, Peter Mancall, Jack Matthews, Drew Newman, Lindsay O'Neill, Sharon Oster, Dan Richter, Roy Ritchie, Jason Sharples, Cynthia Smith, and Stefanie Sobelle.

Opportunities to present my work at different stages between 2014 and 2021 significantly improved this book. I particularly thank fellow panelists and audience members at conferences hosted by the following professional societies: the Society of Nineteenth-Century Americanists, the Society of Early Americanists, the Association for the Study of Literature and Environment, the American Comparative Literature Association, and the Society for Historians of the Early American Republic. I am also grateful to UCLA's Americanist Research Colloquium and particularly to Chris Looby and Carrie Hyde for organizing it, to the Library Company of Philadelphia's Davida T. Deutsch Program in Women's History for inviting me to give the 2021 Annual Women's History Month Lecture, to the Miami University Department of English Work-in-Progress Series, and to the Newberry Library's podcast series *Shelf Life*, particularly Keelin Burke for arranging an episode on my research.

At Princeton University Press I am grateful to Anne Savarese for her initial enthusiasm for the project, her wise editorial guidance, and her support throughout the process of publishing a book in a pandemic. At Princeton I also thank James Collier and Natalie Baan for their expert assistance with manuscript preparation. And I am grateful to the anonymous external readers whose suggestions greatly improved the book.

While writing this book I learned that coral brings people together: it starts conversations not only about the tragedy of coral reef loss but also about a grandmother's necklace, a family's collection of coral specimens, or a visit to the aquarium. Thank you to everyone who was willing to "talk coral" with me. Though it is not possible to thank each person individually, please know that each one of our conversations deepened my understanding.

A lively network of friends continues to sustain me and make my work possible. Thank you to the original Florida crew, Tracie Ashe, Nick Carrabba, Cristi Kupstas, and Carol and Julian Rice; to Melissa Shim; and to graduate school friends whose friendship becomes dearer over time, particularly Katrina Harack, Arden Stern, Erin Walsh and Paul Pender, Aaron Winter and Susannah Rosenblatt, Marcello Giuliano, and Christine Connell Giuliano, for the endless research library assistance. This profession is still unimaginable without the friendship of Cynthia Nazarian and Stefan Vander Elst, Christina Snyder and Jacob Lee, Scott Sowerby and Wayne Huang, Lindsay Schakenbach Regele and Matt Regele, Lara and Tim Crowley, and Elizabeth Stockton and Ryan Brown. Audrey Wasser and Rob Lehman, your backyard, for warm drinks and the best conversations on cold nights, has been our Oxford survival pod and so much more; our lives are immeasurably better for including you. Ian and Kristine Crockett: thank you for being a central, fundamental part of why Los Angeles always feels like home. Michelle Neely and Dominic Mastroianni remain the best book club that ever there was. I owe both of them boundless gratitude for sharing their brilliant minds, love and friendship, laughter and absurdity, understanding and acceptance beyond all reason, and for dance parties yet to be. And all my friends' children, younger and older, too many to list, have my immeasurable gratitude just for being.

My family continues to be a significant source of strength and stability, beginning with my mother, Dianne Currie. Mom, thanks beyond words for your love, listening, and encouragement, and your remarkable capacity to adapt to all that life brings, a quality that I imagine you got partly from your own mother, Elaine Munyer. For surrounding me with love and support, I am grateful to Sharon and Bill Conley; Will Conley and Ivy Wang; Dianne, Mike, Lily, Arden, and Conley Del Bueno; Jeannine, E. J., Dylan, and Noah Shalaby; and the one-of-a-kind Marilyn Munyer, who lives on in us through her dauntless sense of adventure and love of all things coastal California. I also could not imagine more extraordinary parents-in-law than Francine and Edward Navakas, whose love, humor, compassion, and wide-ranging knowledge of literature and life continue to sustain and expand our world. I am likewise grateful to various delightful people across the rest of the Navakas family (Gabe, Lily, Isabella, Sam, and Eliana) and the Glasberg family (Myrna, Rich, Jade, and Burton) and thankful to have known and loved Annette Glasberg and Adele Ginsburg. My Dad, Mitch Currie, aka Screen Man, did not live to see this book's publication. But he cheered me on through much of its research and writing, not least by teaching me that one can always find some way to keep rockin' 'n' rollin'.

This book is dedicated to Gene Navakas. Thank you for listening, and for really hearing every word. Thank you for sharing life with me, wherever it takes us. Thanks most of all for showing me that, even in the apparently familiar, there is always more to see and more occasion to wonder. For all this brilliance, and much more, there could be no better mind and heart.

Whatever thankfulness I have left is for the trees sustained by the Great Miami River and the Arroyo Seco, and every bird that sings in them.

*Pasadena, CA, July 2022*

Portions of this manuscript have previously appeared as essays in *American Literature* ("Antebellum Coral") and *Age of Revolutions* ("Coral, Labor, Slavery, and Silence in the Archives"). I am grateful to both venues for permission to reproduce.

# NOTES

## INTRODUCTION

1. Walls, *Seeing New Worlds*, 5–6. Darwin's theory of reef formation, as expressed in *The Structure and Distribution of Coral Reefs* (1842), is that reefs grow only in places where the seafloor sinks slowly enough for coral polyps to build upward so as to remain in the shallow, warm waters they require. Lyell conceded the general correctness of the theory, which supported the overarching position that landforms emerge slowly and gradually, rather than through sudden alterations. For an excellent discussion of this debate, see Rudwick, *Worlds before Adam*, 489–92. Darwin's coral theory was widely discussed and excerpted in contemporary US periodicals, including *Godey's Lady's Book*. Hale, "Coral Reefs," 240. For more on the popular uptake of Darwin's coral science, see chapter 3.

2. Coral polyps are a species of colonial invertebrates, among which reef-building polyps are only one kind. These polyps live in shallow, tropical waters and filter seawater through their bodies, which contain microscopic algae that photosynthesize to convert the water's nutrients into the polyp's food. As polyps grow, they deposit a hard exoskeleton, which is what we see as a reef. Reefs, then, are a consortium of many organisms: they emerge through a symbiotic relationship among animal polyps, microscopic algae, and multiple bacteria and viruses. I base my definition of reef-building corals largely on the most accessible and accurate definition of coral that I have found, that offered by late coral scientist Dr. Ruth Gates around minute 12:30 in the 2017 Netflix documentary *Chasing Coral* by Jeff Orlowski-Yang. For more on the popular uptake of biology, geology, and other natural sciences prior to the late nineteenth century, when these and other disciplines became more specialized and professionalized, see Rusert, *Fugitive Science*, 5, 15–22. For more on popular and scientific understandings of coral reef formation during the nineteenth century, see Sponsel, *Darwin's Evolving Identity*, 19–24. For an accessible discussion of the present state of coral science as of 2018, see Braverman, *Coral Whisperers*, 6. See also James Bowen's history of coral reef science. Bowen, *Coral Reef Era*.

3. Hale, "Coral Branch," 5.

4. Marx, *Capital*, 339.

5. Marx, 327.

6. Sigourney, "The Coral Insect." This poem is reproduced in full, with additional analysis and discussion of publication history, in chapter 2.

7. Sigourney, "The Coral Insect."

8. The term "Americans" of course, includes peoples far beyond US borders. For the sake of brevity and conciseness, however, in this study I generally use the term "America" to refer to the US, and "Americans" to refer to those living within US borders.

9. The term "capitalism" was not used in print until the 1850s and did not become associated with Marx until the publication of *Das Kapital* in 1867. Furthermore, neither "industrial" nor "capitalism" accurately describe an early nineteenth-century US economy that was still largely agrarian. Schakenbach Regele, *Manufacturing Advantage*, 4. Nonetheless, "capitalism as a system of economic behaviors and relations" long predates the popular emergence of capitalism as a concept, and it is this system of behaviors and relations to which I refer. Schakenbach Regele, "A Brief History," 2–5. My use of the term "slavery's capitalism" borrows from the title of a collection of essays edited by Sven Beckert and Seth Rockman that explores various histories of "racial capitalism," a term first used by Cedric Robinson to challenge a more traditionally Marxist account of capitalism that overly emphasizes capitalism's reliance on wage labor. As Walter Johnson explains it, "In actual historical fact there was no nineteenth-century capitalism without slavery." Johnson, *River of Dark Dreams*, 254. As

Edward Baptist shows, "The expansion of slavery in many ways shaped the story of *everything* in the pre–Civil War United States." Baptist, *Half Has Never*, xxii. Slavery aided the rise of industrial capitalism both because of "the actual work that enslaved men and women performed" and because of slavery's "'cultural work.'" Rockman, *Scraping By*, 7. As Nikhil Pal Singh explains, drawing on Beckert, though slavery "preceded industrialization," slavery's "technical, intellectual, and economic legacies were key to capitalism's expansion"; in particular, slavery installed "forms of both coercion and ethical and political devaluation" that aided capitalist growth. Singh, *Race*, 79, 81. One way of explaining this is that slavery radiated outward to aid the rapid rise of other, very different, forms of coercive labor, including two that this book discusses, wage labor and the unpaid reproductive and domestic work of those gendered as women.

It is important to note at the outset of this study that these other forms of labor are in no way equivalent to slavery. Enslaved persons were the legal property of their enslavers, and so the "social death" of slavery (Patterson, *Slavery and Social Death*, 35–77) is an experience distinct from the exploitation experienced by wage workers, however much the latter are constrained by the forces that Rockman identifies, including "law, economic necessity, and the superior power of employers." Rockman, *Scraping By*, 258. And however constraining the subordination and domestic servitude experienced by housewives, these women did not, by and large, as Angela Davis observes, experience the routine "whips and chains," "floggings and rape," that shaped the lives of countless enslaved Black women. Davis, *Women, Race and Class*, 34, 27. At the same time, we must also consider many forms of unfree or marginal labor alongside one another in order to leverage a truly anticapitalist critique (Singh, *Race*, 88; and Beckert and Rockman, *Slavery's Capitalism*, 9). Doing so allows us to perceive most clearly that "freedom" under capitalism is indebted to slavery; that the precondition of capitalism is coercion, whether commodification, lack of access to power, or compulsion imposed by law, custom, or necessity; and that shared ideologies emerged to rationalize the human hierarchies created and required by coercive labor of different kinds.

10. I will have more to say on the definition and rise of biological essentialism and scientific racism below and in chapter 3.

11. Formative studies of the variegated and complex forms taken by the early US political imagination of coercive labor, as that reality conflicted with the ideal of a society dedicated to unprecedented freedom, include Rockman, *Scraping By*; Boydston, *Home and Work*; Roediger, *Wages of Whiteness*; Nelson, *Commons Democracy*; Dillon, *Gender of Freedom*; Merish, *Archives of Labor*; and Hart, *Trading Spaces*. Yet these studies focus on more explicit discussions of human labor, and (perhaps consequently) leave little room to imagine that those who benefited most from capitalist labor relations— and their attendant ideal of the liberal subject as autonomous, male, and heteronormative—could also acknowledge, mourn, or critique those relations. Rather, such scholarship tends to suggest one of two possibilities. On the one hand, people erased or minimized coercion: on the "social fictions" of "free" labor relations peddled by factory owners and politicians, see Rockman, *Scraping By*, 4; on "fictions of liberalism" that "[humanize] the world through the erasure of profit-generating labor," and particularly women's domestic and reproductive work (which is fashioned as "no labor–only love" rather than toil performed under "constraint" and out of "necessity or compulsion"), see Dillon, *Gender of Freedom*, 202, 201; on how motherhood came to be "imagined as solely emotional and spiritual," see Doyle, *Maternal Bodies*, 21; on "the pastoralization of house work" and how women themselves came to doubt and deny the value of their labor, see Boydston, *Home and Work*, 142–63, xii; on whiteness as a "wage" accepted by white workers "to make up for alienating and exploitative class relationships" that they experienced, see Roediger, *Wages of Whiteness*, 13; and on overt celebrations of a form of "commons democracy" that, for all its emphasis on the value of the collective, "was not anti-capitalist," see Nelson, *Commons Democracy*, 10. On the other hand, people (usually laborers and their allies) demanded social and economic reform: on labor movements, unions, and strikes, see Rockman, *Scraping By*, 148–52; on working-class women writers of many racial identities who forged new solidarities while protesting exploitative labor conditions and demanding economic justice, see Merish, *Archives of Labor*; and on "the emergence of abolitionist mass media," see Goddu, *Selling Antislavery*, 1. My study both builds on and challenges that relative division (downplaying coercion vs. demanding reform) by locating much less explicit, widespread ways that people of various

personal identities, political affiliations, and social positions acknowledged, and even normalized, coercion through coral.

12. As Lorraine Daston observes in groundbreaking work on the relation between nonhuman nature and human society, nature's "authority derives from the way it eases thinking about otherwise intractable problems. One such problem is how to imagine the structure and functioning of society, an entity at once all-pervasive but invisible." Daston, *Moral Authority*, 9.

13. For a trailblazing study of some more socially just models of nineteenth-century environmentalism in particular, see the work of Michelle C. Neely, who locates some of these models in earlier US writing and reflects on the urgency of putting them into practice now. Neely, *Against Sustainability*.

14. I draw this "biographical approach" from Igor Kopytoff, and more particularly from the way that Akinwumi Ogundiran employs and modifies Kopytoff's methods to uncover the cultural biography of cowrie. Kopytoff, "Cultural Biography," 66–68; and Ogundiran, "Of Small Things Remembered." And although coral, as an object, differs from the manufactured objects of later US writing that are the subject of Bill Brown's landmark study of "the object matter of American literature," I am deeply influenced by Brown's reflections on "the complex roles that objects have played in American lives." Brown, *Sense of Things*, 12. I will have more to say about my particular approach to coral as material and metaphor in chapter 1.

15. In that aim I have been especially guided by Susan Scott Parrish's discussion of "rummaging" and her particular approach of "following the animal," as she describes it, drawing on George Marcus's description of "'multi-sited research . . . designed around chains, paths, threads, conjunctions, or juxtapositions of locations'." Parrish, "Rummaging," 268, 266–67. Parrish practices this method of wide-ranging research across many media in her book, *American Curiosity*, an important and early inspiration for all of my work.

16. As Linda Tuhiwai Smith reminds us, the very "history of Western research" is "deeply embedded in . . . multiple layers of imperial and colonial practices" that are difficult to avoid reproducing even today. Smith, *Decolonizing*, 2. As Destin Jenkins and Justin Leroy explain, drawing on Saidiya Hartman's formative discussion of the "afterlife of slavery," "the temporality of racial capitalism is one of ongoingness, even if its precise nature is dynamic and changing. It is a process, not a moment." Jenkins and Leroy, *Histories of Racial Capitalism*, 12. Hartman's phrase refers to the multiple and intersecting economic and cultural legacies of slavery that Black people continue to face in the form of "skewed life chances, limited access to health and education, premature death, incarceration, and impoverishment." Hartman, *Lose Your Mother*, 6.

17. For studies of how European and American naturalists relied on the knowledge, labor, and skills of local, frequently marginalized populations for specimen collection and analysis, see Strang, *Frontiers of Science*, and Parrish, *American Curiosity*. Parrish reflects as follows on the inseparability of natural history and imperialism: "As the very idea of a specimen implies the splitting into parts a replete realm of nature, as the taxonomy of nature was implicated in the taxonomy of human beings and sexes that ultimately favored the white male arbiter of categories, and as many naturalists participated in both mercantilism and, in many cases, the owning of slaves, we must see these two realms, not as distinct, but as mutually instantiating points on a continuum." Parrish, *American Curiosity*, 173. Recent and generative studies of the relationship between natural history and imperialism include LaFleur, *Natural History of Sexuality*; Delbourgo, *Collecting the World*; Iannini, *Fatal Revolutions*; and Warsh, *American Baroque*. For the indebtedness of eighteenth-century naturalist knowledge to colonial violence, and to sexual violence against Black women in particular, see Polcha, "Voyeur."

18. By nineteenth-century biological essentialism, I mean "a sense of man as primarily a biological being, embedded in nature and governed by biological laws," and the accompanying idea that "race is . . . imagined to be somehow perceptible—often via visualization." Stepan, *Idea of Race*, 4; and Fielder *Relative Races*, 11. Some especially useful accounts of biological essentialism, as alternately manifested and contested within nineteenth-century US racial science and literary representation, are offered by Dain, *A Hideous Monster*; Rusert, *Fugitive Science*; Ellis, *Antebellum Posthuman*; and Fielder, *Relative Races*.

19. On how and why pre-1900 conceptions of human variety were never a matter of skin color alone, see Wheeler, *Complexion of Race*; Bauer and Mazzotti, *Creole Subjects*; Curran, *Anatomy of Blackness*;

and LaFleur, *Natural History of Sexuality*. On the many economic, cultural, and political transformations that drove the rise of the nineteenth-century life sciences and produced a strictly biological account of race under pressure to rationalize slavery, see Jordan, *White over Black* and Fredrickson, *Black Image*. Colette Guillaumin, Robyn Wiegman, and Colin Dayan offer especially useful accounts of how and why nineteenth-century practitioners of racial science warped earlier and more fluid and arbitrary taxonomies (by Linnaeus and others) into fixed, determinative, and hierarchical schemes. See Guillaumin, *Racism, Sexism*; Wiegman, *American Anatomies*; and Dayan, *The Law*. I will return to this topic in chapter 3.

20. I will take up these racialized ways of seeing and evaluating at length in chapter 3.

21. Here I want to acknowledge that the available language for discussing human racial and gender differences is inevitably problematic and limiting, and it often risks reproducing biological essentialism, even when we mean to interrogate or dismantle it. Following Kimberlé Williams Crenshaw and others, I capitalize Black throughout as a way to refer to persons of African descent as "a specific cultural group" requiring denotation as a proper noun. Crenshaw, "Race, Reform, and Retrenchment," *Race*, 1332. In cases where it seems especially important to highlight the cultural (as opposed to biological) nature of race and/or gender, I use phrases such as "persons considered Black according to the codes of the period" or "persons gendered as women." When, for clarity or brevity, I use terms such as "woman" or "Black person," I never intend these terms as biological descriptors, but rather as signals that such persons are racialized or gendered (or both) in relation to whiteness or maleness (or both). An especially insightful, recent discussion of racial and racialized terminology is that of Brigitte Fielder, who encourages us to acknowledge that it is not possible to fully escape "the limitations of dominant, historical, essentialist racial discourse for discussing race's complexity, construction, and shifting terrain," even as we must try. Fielder, *Relative Races*, 9.

22. Weheliye, *Habeas Viscus*, 40.

23. Barrett, "Identities and Identity Studies," 186.

24. Barrett, 293.

25. For the most succinct formulation of this claim that I know, see Kandice Chuh, who argues for a "refusal of 'aboutness'" since to claim that this or that text is "about" racial identity, while another is definitively not, is "a silo mentality that holds knowledge formations apart from each other" in the aim of upholding the essentialist idea that there is such a thing as discrete identities that precede the political, cultural, economic, and other realities and discourses that form these identities. Chuh, "Not about Anything," 132. Put another way, aboutness can be a form of racism and sexism, a point that the work of many scholars working in Black studies and Black feminist studies also make quite persuasively, including Lindon Barrett and Alexander G. Weheliye, and that Joan Scott lucidly explains in relation to gender and race. As Michel-Rolph Trouillot powerfully reminds us, "there is no perfect closure of any event, however one chooses to define the boundaries of that event"; this is particularly so in the case of a world-making, world-ending event that involved countless millions. Trouillot, *Silencing the Past*, 49. My thanks to Cristobal Silva for first pointing me toward Trouillot's work, and for sharing his own writing and thoughts on Trouillot.

26. McKittrick, *Dear Science*, 135.

27. Bernstein, *Racial Innocence*, 8–13; 69–91.

28. Bernstein, 12.

29. Ogundiran and Saunders, *Materialities of Ritual*, 1–2.

30. LaCapra, *Understanding Others*, 155, 156, 167.

31. Jennifer Fleissner's reflections on LaCapra's analysis of these prospects are particularly useful in this regard. As Fleissner explains, to read the past as inextricable from the present is to refuse a triumphalist arc whereby "the past exists in order to irradiate the virtues of the present"; it is instead to approach the past *not* from a safe, "self-satisfied," more knowing distance but rather as incomplete, ongoing, and inevitably "freighted with the weight of history." Fleissner, "Historicism Blues," 702–3.

32. I echo the title of Janet Neary's recently edited volume of Lindon Barrett's work, *Conditions of the Present*, which Neary in turn borrows from Wiegman's essay in that volume. Within that context,

and as I use it here, the phrase signals the possibilities of scholarly criticism to "register and rework the epistemological conditions of the present," and particularly those conditions forged in the past that continue to install the ideology of racial essentialism in the service of dispossessing nonwhite persons. Barrett, *Conditions of the Present*, 7; 275.

## CHAPTER 1: THE GLOBAL BIOGRAPHY OF EARLY AMERICAN CORAL

1. A biographical approach, as Kopytoff explains, enables us to ask questions of an object that are similar to those we might ask of people, such as, Where does it come from? How was it made (and by whom or what)? What (and where) has been its career thus far? And how and why has it changed over time? Kopytoff, "Cultural Biography of Things," 66. This approach, moreover, considers how and why an object gains "culturally specific meanings" as it is "redefined and put to use" among different cultures and peoples over time. Kopytoff, "Cultural Biography of Things," 68, 67. The objective of a cultural biography is to reconstruct how an object and its meanings emerge from multiple, changing, and intersecting contexts, whether economic, political, familial, scientific, social, or otherwise. Crucially, this approach remembers that an object acquires meaning through and because of its "genealogies of practice." Ogundiran and Saunders, *Materialities*, 20–22. In other words, the meaning of an object in any culture derives from much longer histories of the object's uses and meanings in multiple cultures. To the extent that it is possible, then, I have tried to reconstruct several genealogies of coral, while also documenting early American awareness of them; at the same time, however, I am guided by Robin Bernstein's observation that things may prompt actions and responses—things may "script a repertoire"—whose meaning exceeds the knowledge or intention of any given actor or author. Bernstein, *Racial Innocence*, 19, 12. We must therefore also ask, "'What historically located behaviors did this artifact invite? And what practices did it discourage?'" Bernstein, 8. And I have kept these genealogies in mind when addressing the questions that Brown poses of "the object matter of American literature," including "What desires did objects organize? What fantasies did they provoke? Through what economies were they assigned new value? Through what epistemologies were they assigned meaning?" Brown, *Sense of Things*, 12.

   The word *coral* probably derives from Greek. Diderot's widely read *Encyclopedia* (1754) records that "Coral, according to some, draws its name from the Greek words κόρειν, to adorn, and ἁλός, sea, as if there were no other marine production whose beauty could be compared to coral: and there is no aspect of it on which neither the ancients nor the moderns have written much." Diderot, *Encyclopedia of Diderot*, 4:194–96.

2. Studies that approach coral more globally, and that have proven particularly valuable to this book in that regard, include Vermeren, "Être corailleur en Algérie"; Buti, "Du rouge"; Calcagno, "A caccia"; Lo Basso and Raveaux, "Introduction"; Liverino, *Red Coral*; Yogev, *Coral and Diamonds*; Trivellato, *Familiarity of Strangers*, 224–50; Endt-Jones, "Monstrous Transformation"; and Kelley, "King's Coral Body." Coral is important in cultures that do not feature in this study because I did not detect the influence of these cultures on the early US conceptions of coral that are the focus of this study. For example, as Yogev documents, an enormous amount of coral harvested in the Mediterranean was traded through London (and frequently through Jewish merchants there) to India, where it was exchanged for diamonds and used for jewelry, funeral practices, and as a symbol of social status. Yogev, *Coral and Diamonds*, 102–3. Trivellato documents the role of Jewish merchants at Livorno in this trade in *Familiarity of Strangers*, 225–38. Coral also holds long-standing cultural and economic importance in parts of Latin America; and the same is true in Japan and China, as Nozomu shows in an analysis of the "coral road" in *Precious Coral*. And elsewhere in the Pacific, coral holds great religious, symbolic, and material significance; in that connection, the frequent appearance of coral in the works of Melville could be a study all on its own. I should have liked to study coral's meaning within Indigenous cultures in North America; during the period under consideration, I could not find reliable evidence of the importance of coral within these cultures, though this could be due to

the colonial violence that destroyed or displaced Indigenous populations, preventing the formation of literary, oral, or material traditions of coral.

3. Joshua Johnson (1795–1824) was largely self-taught, though he was highly conversant in the works of artists in the Peale family with whom he may have closely associated, according to Weekley et al., *Joshua Johnson*, 50–54. For additional biographical information on Johnson, see Weekley et al., *Joshua Johnson*, 47–64; and Isaacs, "Joshua Johnson," 59–60. Johnson's portrait of Emma Van Name—a name that Weekley suggests might be a variation of the Baltimore family surname "Van Noemer"—is considered to be "his most ambitious and engaging portrait of an individual child" according to the archival note by the *Metropolitan Museum of Art*. "Emma Van Name, ca. 1805," Metropolitan Museum of Art (website), accessed September 24, 2022, https://www.metmuseum .org/art/collection/search/701989. Weekley et al., *Joshua Johnson*, 129. The portrait also features the most coral of any of his portraits that I have seen. Additional examples of portraits by Johnson featuring coral jewelry include (but are not limited to): *Mrs. Kennedy Long and Her Children* (1805, private collection; it also features a coral and bells); *Elizabeth Grant Bankson Beatty (Mrs. James Beatty) and Her Daughter Susan* (1805, Art Institute of Chicago); *Adelia Ellender* (1803–5, Smithsonian American Art Museum); *Grace Allison McCurdy (Mrs. Hugh McCurdy) and Her Daughters, Mary Jane and Letitia Grace* (1806, Corcoran Gallery of Art); *Elizabeth (Mrs. Andrew) Aitkin and her Daughter Eliza* (1805, Museum of Fine Arts-Boston); *Woman and Baby Wearing Green Gloves* (1800, private collection); and *Seated Girl with Strawberries* (private collection).

4. On popular varieties of nineteenth-century coral jewelry in England in particular, see Anderson, "Coral Jewellery," 47–48. Torntore offers illustrations and descriptions of red coral bead varieties. Torntore, "Precious Red Corals," 8. Some examples of early American portraits of women, girls, and babies of all genders in coral include Thomas Sully, *Mrs. James Montgomery, Jr.* (1845, Metropolitan Museum of Art); Charles Willson Peale, *Mrs. Laurent Clerc* (1822, Wadsworth Atheneaum Museum of Art); John Bradley, *Emma Homan* (1844, Metropolitan Museum of Art); Orlando Hand Bears, *Miss Tweedy of Brooklyn* (1845, Detroit Institute of Arts); Clarissa Peters, *Baby with Rattle and Dog* (1842, Metropolitan Museum of Art); Frederick R. Spencer, *Mary Ann Garrits* (1834, Metropolitan Museum of Art ); and George Caleb Bingham, *Mrs. B. W. Clark (Mary Jane Kinney) and Her Brother (Joseph Beeler Kinney?)* (1874, Nelson-Atkins Museum of Art). Many unknown painters working in the folk art tradition also portrayed their subjects in coral. Many museums hold coral tiaras worn by nineteenth-century American women, but I have not seen these items depicted in early American portraits.

5. Hale, "Coral Reefs," 239.

6. Hale, 240. Part of this description is drawn from Flinders, *Terra Australis* (1814).

7. Hale, "Coral Reefs," 239. According to Anderson, Victorian coral ornaments evoked "Victorian ideas (and anxieties) about marine life and earth's history," and though Anderson does not discuss the topic of labor, I take inspiration from her insight that "a coral object brought [larger scientific and cultural] concerns imaginatively down to earth, where they could be easily stroked by the fingers or held on the palm." Anderson, "Coral Jewellery," 48. For example, Samuel Griswold Goodrich's chapter on coral (127–38) in his popular natural history for children, *Peter Parley's Wonders of the Earth, Sea, and Sky* (1840), connects coral ornaments directly to coral reef science: "You are acquainted with the appearance of coral, as you have often seen pieces of it in cabinets, and employed as ornaments," begins the chapter, which proceeds to describe the science of reef formation. Goodrich, *Peter Parley's Wonders*, 127. The link between a coral necklace, laboring polyps, and laboring coral fishers is made in US writer Margaret Coxe's evangelical pamphlet, *The Wonders of the Deep; or Two Months at the Seashore*. Coxe, 99.

8. Moody published popular works on natural history through an international firm with offices in London and New York. Her work was popular in the US and abroad: *Fairy Tree* was reprinted under its original title in 1872 and multiple times thereafter as *Far and Near; or, Stories of a Christmas Tree* (1864, 1865, 1866, 1867, 1870). The American Antiquarian Society's copy of this latter work was owned by a New England family, as the inscription suggests.

9. Moody, "Coral Bracelet," 30, 9.

10. Moody, 30.

11. Moody, 31.
12. Moody, 31.
13. Moody, 31–32.
14. Moody, 33, 34, 36.
15. Moody, 36.
16. Moody, 34, 36.
17. Moody, 37, 38.
18. Moody, 32.
19. Spallanzani's popular US uptake is indicated by the many excerpts from and discussions of his coral fishing chapter that appear in US periodicals, as well as in popular stories about coral, such as that by Coxe (1836), whose female narrator also cites Spallanzani and draws extensively from his work when teaching her children about coral. Coxe, *Wonders of the Deep*, 97–99.
20. Spallanzani drew his account mainly from interviews with local fishermen, many of whom he accompanied to sea, but also from the much earlier account of Luigi Ferdinando Marsigli, who did the same during the previous century in coastal France. Spallanzani, *Travels*, 315. On Spallanzani's sources see also Vandersmissen, "Experiments," 58. For more on specifically where Mediterranean coral grew and was harvested see Trivellato, *Familiarity of Strangers*, 226–27. The practice of coral fishing has a very long history and has taken different forms in different places, beginning in the Minoan-Mycenaean era (3000 BC–1200 BC). For my description of coral fishing, processing, and trading in this chapter I draw mainly on three sources: Trivellato's brief discussion, which in turn relies on French and Italian historical and scholarly accounts. Trivellato, 225–32; Spallanzani's historical descriptions. Spallanzani, *Travels*, 306–29; and Liverino's descriptions of coral fishing and workmanship, past and present. Liverino, *Red Coral*, 72–91, 103–118. Liverino was a self-taught historian from a family of coral fishers dating to the nineteenth century and the founder of the Museum of Coral at Torre del Greco, a major center of coral fishing and manufacture. For information on how and where the practice of coral fishing continues today in the Mediterranean and elsewhere, a task now performed by divers (*coraleros*), along with the many problems this practice causes, see Rossi, *Oceans in Decline*, 117–27.
21. According to Buti, coral fishers consisted mainly of Corsicans, Catalans, Provençaux, Genoese, Neapolitans and Sicilians who vied with one another for fishing access and rights along the coasts of North Africa. Buti, "Du rouge," 116. The coral fishers, called *corallini* in Italian and *corailleurs* in French, were part of a vast network that also involved boat managers and owners, sailors, merchants, traders, manufacturers, and coral workers who transformed raw coral into cut and polished pieces. Trivellato, *Familiarity of Strangers*, 225–32. See also Vermeren, "Être corailleur en Algérie." At certain times and places that network involved, or was presided over by, major commercial companies that also traded in enslaved persons, as I discuss below.
22. Spallanzani, *Travels*, 317. Early Americans learned of the dangerous and difficult labor of coral fishing through Spallanzani's popular account as well as numerous other widely read contemporary sources, including periodical essays such as "On Coral and the Coral Fishery" (1840) and reference works such as John Barrow's *Dictionarium polygraphicum* (1735) and John Akin's *Encyclopedia* (1798), both of which relate that, should the rope break when hauling up the coral, the fisherman are in danger of drowning. Knight, "On Coral," 80; Barrow, 198; Akin, *Encyclopedia*, 5:443. Contemporary scholars who discuss these hazards include Liverino, *Red Coral*; Calcagno, "A caccia"; Beri, "Corallatori e guerra"; and Lopez, "Coral Fishermen," 195–211.
23. The design of the *ingegno* dates to antiquity; Pliny describes it in book 32, chapter 11, p. 11. See also Liverino, *Red Coral*, 32.
24. Spallanzani, *Travels*, 309–10; and Moody, "Coral Bracelet," 32. Marsigli (1658–1730), sometimes spelled Marsili, is author of *Histoire Physique de la Mer* (1725), published in French under the pen name Louis Ferdinand comte de Marsilli.
25. Spallanzani, *Travels*, 308.
26. A detailed scholarly account of the threat of corsairs (sometimes termed "Barbary pirates") to the coral fishers, and the measures they took to protect against attack during the eighteenth century in particular, is offered by Beri, "Corallatori e guerra." See also Liverino, *Red Coral*, 75; and Calcagno, "A caccia."

27. Moody, "Story of a Coral Bracelet," 31–32.

28. Moody, 33.

29. Contemporary sources describing coral workers as mostly women include Ghidiglia, "L'industria del Corallo," 499; and Simmonds, *Commercial Products*, 442. Women made up the majority of workers at the forty coral workshops at Torre del Greco in 1883. Haughwout, "The Coral Industry," 121. More recently Torntore explains that the time-intensive methods for producing coral beads in particular have remained relatively unchanged over time: the "beads may go through as many as 12 stages of highly labor-intensive production before they are finished" and some of these stages are "completely accomplished by hand by women working in a factory workshop or at home." Torntore, "Precious Red Corals," 4, 5. I have been unable to find much additional reliable information on the identities of coral workers, though there are studies of coral merchants and workshop owners. For example, as Yogev documents, Anglo-Jewish merchants living in London and elsewhere presided over the interlinked diamond and coral trades during the seventeenth and eighteenth centuries. Yogev, *Coral and Diamonds*. Trivellato discusses Livorno in particular, noting that the industry there was launched by Genoese immigrants, though later run by Jewish merchants. Trivellato, *Familiarity of Strangers*, 230.

30. Moody, "Coral Bracelet," 33.

31. Moody, 37.

32. Moody, 31. On the conventional opening of slave narratives with "I was born . . ." see Olney, "'I Was Born'."

33. Jacobs more fully describes the condition of enslaved Black girls as follows: "Before she is twelve years old . . . she will become prematurely knowing in evil things . . . if God has bestowed beauty upon her, it will prove her greatest curse. That which commands admiration in the white woman only hastens the degradation of the female slave." Jacobs, *Incidents*, 38.

34. Wright, *Black Girlhood*, 14. According to Wright's analysis of this trope, "black writers used black girls as tools to put forward their social and political agendas." Wright, 12. Moody, a white writer whom Wright does not analyze, seems to be participating in this trope.

35. On the prospect that the subjects in this portrait are free, rather than enslaved, and on the allegorical nature of this ethnographic representation, which was part of a series that Eckhout produced, see Brienen, "Albert Eckhout's." Brienen usefully reminds us to avoid reading colonial sources such as this one as documentary; the subjects portrayed in them may be a composite of the artist's observations of many such persons, or these subjects may have been forced to pose for the artist. Brienen, 230–31. Either way, the subjects represented in these paintings did not have control over their representation. Brienen, 236.

36. Porcelain figures of the four continents were produced by major porcelain factories throughout Europe during the eighteenth century, including Derby and Bow (England), Meissen (Germany), and Sèvres (France). Many English porcelain wares in particular were exported to the West Indies and mainland American colonies. Young, *English Porcelain*, 180. Graham Hood documents the use of these figures in stylish colonial Virginia households, noting that the pieces typically ornamented mantelpieces and were moved to the dining table as centerpieces for the dessert course. Hood, *Governor's Palace*, 128–29, 131–32. For information on the use and consumption of these figures in England, see Young, *English Porcelain*, 184–89.

37. The identity of the painter of the Bingham portrait of Topsy and Eva remains unknown. Is "J. A. Bingham" possibly John Armor Bingham, the Ohio abolitionist and Republican politician?

38. Ripa, *Iconologia*, 53. On the representational tradition of the four continents, and "Africa" wearing or holding coral as part of this commercial allegory, see Le Corbeiller, "Miss America," 218; Hughes, "Four Continents," 742; and Spicer, "Personification of Africa," 704.

39. To my knowledge no scholar has suggested the possibility of a connection between this practice of personal ornamentation among communities within Africa, and the Western visual and written convention of a black woman in coral. The possibility of that connection is left open, however, by some scholars of the "four continents" allegorical tradition: Hughes, for example, notes that the convention, in its later iterations, may record not only each continent's "contributions to the global marketplace" but also "the distinct cultural practices" of people in certain regions, as

these practices are recorded in contemporary travel narratives and other accounts. Hughes, "Four Continents," 716.

40.  Two especially well-known travel narratives from this period that document the practice of wearing red coral in various places along the western coast of Africa include Bosman, *Description of Guinea*, 436, 440; and Lander, *Course and Termination of the Niger*. Lander records that "the demand for coral has been very great in every town of consequence which we have visited. All ranks of people appear passionately fond of wearing it, and it is preferred to every other ornament whatever." Lander, 4–5.

41.  Grandpré, *Voyage*, 71. Grandpré records that "l'habillement des femmes est moins noble; . . . elles se . . . au surplus, ainsi ques les hommes, tres-avides de corail; cette substance est pour eux le nec plus ultra de la richesse; elle est a leurs yeux ce que le diamant est pour nous." Grandpré, 74–75. I believe it is safe to assume that some or all of the beads in the image of this woman are red coral, particularly because the same type of beads appear in the accompanying image of a man in the section of Grandpré's text entitled "Costume" that relates, "Ils sont excessivement avides de corail rouge; cest le comble du luxe, et ils le recherchent avec ardeur pour leur parure." Grandpré, 71. Gilbert Buti has recently interpreted the beads as red coral. Buti, "Du rouge," 111.

42.  Ogundiran, "Of Small Things," 431.

43.  Ogundiran, 435.

44.  On the uncertain origins of red coral beads at Yoruba in particular, Ogundiran writes that "the red coral bead" may have preceded the Atlantic economy (via overland transport through the Sahara), but it certainly "became popular following the advent of transatlantic commerce." Ogundiran, "Of Small Things," 432–33. On the oral tradition of Olokun and its emergence in tandem with the arrival of Portuguese slave traders bearing coral, I have drawn on the accounts of Ben-Amos, *Benin*; Blackmun "Iwebo," 27–29; and Welsh, 49–52. On the power of "àṣẹ" held by some of these objects, see Abiodun, *Yoruba Art*, 53–87.

45.  Ben-Amos, *Benin*, 92.

46.  Ben-Amos, 100.

47.  I have arrived at this list of locations by triangulating extant scholarship with passages that I have located in early modern texts (many discussed below) that explicitly describe the exchange of red coral for persons (as opposed to the use of red coral as currency within any given culture more generally). However, the work of documenting the history and precise locations of this trade practice is difficult for a number of reasons. European ethnographic texts are not straightforward documentary sources of information about the local cultures of Africa. Place names in these texts have changed over time. Only a small number of scholars have begun to investigate this topic, and many are working in languages other than English. The most comprehensive and reliable scholarly work on this topic is that of Gilbert Buti, who explains that European traders used coral to initiate contact with local dignitaries and other authorities and for bartering in larger quantities for persons, from roughly the fifteenth to eighteenth centuries (though I have located some sources suggesting the practice continued into the early nineteenth century). Buti, "Du rouge," 125–26. According to Buti, most of this coral came from the western basin of the Mediterranean Sea (primarily along the coasts of Tunisia, Algeria, and France), and many of the fishers harvesting coral there were French and Italian men hired directly by powerful European mercantile companies that facilitated the slave trade, including Great Britain's East India Company and Royal African Company. Buti, 116–17. The latter company shipped more enslaved persons from Africa to the Americas than any other institution in the history of the trade. Pettigrew, *Freedom's Debt*, 11. For passing scholarly mentions of the exchange of coral for persons, see Ryder, *Benin*, 56; Hartman, *Lose Your Mother*, 68; Trivellato, *Familiarity of Strangers*, 226; Hancock, *Citizens*, 190; and Gikandi, *Slavery*, 80.

48.  Dapper, *Naukeurige Beschrijvingen*, 491. The international influence of Dapper's *Naukeurige Beschrijvingen der Afrikaensche gewesten* (1668), often translated in English as *Description of Africa*, would be difficult to overstate. After publication the volume was quickly translated into many languages and became the basis for British writer John Ogilby's magisterial and widely known *Africa* (1670). Dapper evidently based his text on records, now lost, of the Dutch West India Company. The various translations and adaptations of his text do not always include the passages documenting red coral as currency for persons. My thanks to Floris Winckel, a PhD student at the Rachel Carson

Center, Ludwig Maximilian University of Munich, for assistance in translating the relevant passages in Dapper's text.

49. Phillips, *Journal*, 227.
50. Chambon, *Guide du Commerce*, 385.
51. Landolphe, *Mémoires*, 125–29.
52. Conneau, *Adventures*, 347. Additional texts that explicitly reference the trade of coral for persons include Jamieson, *Commerce*, 46; and Des Bruslons, *Dictionnaire*, 372, 374. Coral is discussed as a trade good in some letters of the Royal African Company. Law, *English in West Africa*, 1:4, 198, 246. My thanks to Christopher M. Blakley for pointing me toward trade dictionaries and letters as a valuable source of information on this topic.
53. Buckridge cites and mentions a number of these texts, to which I will return in chapter 5. *Language of Dress*, 58, 86.
54. Also see chapter 5.
55. See the introduction to this book for a discussion of the importance of considering these different groups in relation to one another, without suggesting any equivalence in their experiences of capitalism's multiple forms of coercion.
56. Moody, "Coral Bracelet," 34.
57. Longfellow, *Belfry*, 51–61, ll. 15–25. Information on the Longfellow family's two coral and bells is drawn from the archival note on this item by the Longfellow House, Cambridge, Massachusetts. (See https://www.nps.gov/long/blogs/coral-and-silver-baby-rattle.htm). My thanks to Christopher Allison for mentioning the Longfellow coral and discussing potential associations between coral and familial bloodlines that I develop in this section of the chapter.
58. Ads for the coral and bells in the newspapers listed here (and many other newspapers) are cited and discussed by Katharine Morrison McClinton, who also considers mentions of the object in account books and the relatively high demand for an item of this expense. McClinton, *Antiques of American Childhood*, 24–26. Harry Bischoff Weiss similarly discusses early American newspaper ads along with many early American examples of the object. Weiss, *American Baby Rattles*, 13–15. And Berenice Ball discusses American newspapers, account books, and the other types of sources listed here. Ball, "Whistles," 554–55. My thanks to curator Alexandra Ward for pointing me toward many useful sources on the coral and bells.
59. Beth Carver Wees and Medill Higgins Harvey discuss both locally produced and imported examples, as does Katharine Morrison McClinton. Wees and Harvey, *Early American Silver*, 266; and McClinton, *Antiques of American Childhood*, 26. The National Museum of American History holds a coral and bells attributed to Louisa Catherine Adams, first lady and wife of John Quincy Adams.
60. The medicinal and apotropaic properties of coral are discussed in many sources during the period. In different versions of *Encyclopedia* Diderot discusses coral as medicine and as apotropaic, according to Gilbert Buti. Diderot, *Encyclopedia*, 4: 196–97; and Buti, "Du rouge," 110. For nineteenth-century sources see Barrera, *Gems and Jewels*, 250–51; and Nimmo, *Omens*, 59. Scholarship on these properties as the primary reason that babies and children wore coral (in life and art) includes Callisen, "Evil Eye"; Calvert, *Children*, 48–49; Hansen, "Coral," 1424; and Endt-Jones, *Coral*, 12–13. On the object's popularity as a christening present, see Wees and Harvey, *Early American Silver*, 265; and McClinton, *Antiques of American Childhood*, 24.
61. Pliny, bk. 32, ch. 11, p. 11.
62. In *The Boke of Chyldren* (1544), a Renaissance tract on pediatric medicine, for example, Thomas Phaer recommends "redde coral" as protection against epilepsy. Phaer, *Boke of Chyldren*, 41.
63. Ovid, *Metamorphoses*, bk. 4, 706–752.
64. Pliny calls red coral "gorgonia" after the Ovidian myth and to emphasize that it hardens at contact with air. Book 37, ch. 59, p. 450. The misconception is repeated by numerous later naturalists including Robert Boyle. Boyle, *Origine*, 88.
65. "Cousin Annie's Riddle," 12.
66. Watson, "The Coral and Bell," 24. This text purports to offer a pat moral in the closing lines: coral shows us we can all accomplish a lot if we work together. Yet the actual examples of labor that pro-

duced the coral object belie that romantic claim by describing those who toil unseen "with labor and with pain" for others to thrive.

67. No scholar that I am aware of discusses the coral and bells as a fertility object, mentions its phallic shape, or points out that its placement in familial portraiture highlights the specific connections to women's reproduction and/or familial bloodlines. In fact, there has been little discussion of historical links between coral and women at all, though Kelley helpfully observes that "in Ovid's account, coral absorbs its power from female sources," including Medusa, Andromeda, and sea nymphs. Kelley, "King's Coral Body," 124. Certainly, like many other prestige objects, the coral and bells could signify the parents' social standing or wealth. See Wees and Harvey, *Early American Silver*, 265; Calvert, *Children*, 6; McClinton, *Antiques*, 24. Yet my research suggests that many other meanings, specifically gendered meanings related to this particular item, also resonate through these portraits.

68. The origins of the *mano figa* tradition as a charm in multiple cultures is discussed by von Kemnitz, who mentions that it was sometimes given to newborns. Von Kemnitz, "Porous Frontiers," 262–63. The sexual nature and meaning of this object is discussed by John H. Elliott. Elliott, *Beware*, 180–82. My thanks to Wendy Tronrud for calling my attention to this object and its history.

69. Lovell, *Art*, 10.

70. "Ovid, as it happens, did not even mention blood, specifying that seaweed was petrified at Medusa's *touch*." Cole, "Cellini's Blood," 228.

71. Ovid, *Metamorphoses*, qtd. in Cole, 228. The lines appear in the earliest published Italian translation, printed in 1497.

72. For example, Diderot's *Encyclopedia* (1754) records of coral that "mythology considers this plant to have originated from the *blood* of the head of Medusa. It was this monster's final petrification." Diderot, *Encyclopedia*, 4:197, emphasis mine. For one of many early US texts attributing blood to Ovid, see *Encyclopedia Americana* (1833), which states that "as Ovid relates in his *Metamorphoses*," Medusa's blood "tinged" coral red. Ovid, *Metamorphoses*, 530. According to Thorndike, by the mid-seventeenth century coral had attained "a peculiar sympathy with human blood, the signature of red coral seeming exactly to represent the tincture of the nectar of the microcosm." Thorndike, *History of Magic*, 163.

73. Melville's narrator imagines of Ahab that "his whole high, broad form, seemed made of solid bronze, and shaped in an unalterable mould, like Cellini's cast *Perseus*." Melville, *Moby-Dick*, 128. Cole argues that the blood in Cellini's work turns to coral. Cole, "Cellini's Blood," 229.

74. Moody, "Coral Bracelet," 36, emphasis mine. Precisely when *birth* became associated with coral via the Medusa myth is not yet clear to me, though the association certainly shapes early American understandings of coral, and the coral and bells in particular. A number of scholarly treatments of the myth of Medusa in relation to coral erroneously assume or imply that Ovid himself links coral to blood, and much of this scholarship also refers to Ovid's story as "the birth of coral"—without explaining where the "birth" concept comes from. For example, see Torntore on Ovid's Medusa's "blood" as the mythological source of coral, "Precious Red Corals," 11. Torntore here draws upon the work of Basilio Liverino, as translated by Jane Helen Johnson. Liverino, *Red Coral*, 9. See also Sitch and Sugden, "Perseus," 41–44; and Endt-Jones, *Coral*, 9.

75. Lovell, *Art*, 155.

76. Coxe, *Wonders of the Deep*, 7.

77. Coxe, 95.

78. Coxe, 95.

79. Coxe, 94.

80. Coxe, 97–99. Though the titlepage of *Wonders of the Deep* credits only "a lady" as author, I attribute the pamphlet to Coxe because she is the author of *A Visit to Nahant* (1839), the sequel to *Wonders*. Coxe's sources for information on coral in *Wonders*—some cited, others not—also include the entry on coral in *Encyclopedia Americana* (1829); *Good's Book of Nature* (1834); Tyerman's *Journal of Voyages and Travels* (1831); Montgomery's *Pelican Island* (1826); and Turner's *Sacred History of the World* (1832).

81. Anderson and Price, xlii.

82. James L. Reveal identifies Catesby's "Keratophyton fruticis specie, nigrum" as *Plexaura flexuosa*, a species of densely branching soft coral. Reveal, "Identification," 48. On the nineteenth-century popularity of beachcombing for seaweed, sometimes with guides, see Duggins, "Pacific Ocean Flowers." There is sometimes terminological confusion during the period about the difference between "corals" and "corallines," the latter being seaweed.

83. Emerson enthusiastically documents his visit to the gallery, even mentioning the "corals" he saw, in an 1843 letter to his wife. Emerson, *Letters*, 123. For Melville's interest in the publications of the US Exploring Expedition, also known as the Wilkes expedition, see Giles, *Antipodean America*, 142. The specimens of the US Exploring Expedition remained on display in the Patent Office Gallery— currently the Reynolds Center at the Gallery of American Art—from 1843 to 1858. Nathaniel Philbrick describes this public display. Philbrick, *Sea of Glory*, 333–36. For Dana's description of the red coral specimen, see *Zoophyte Atlas*, pl. 60, fig. 1; and *Synopsis of the Report on Zoophytes*, 130.

84. Bellion, *Citizen Spectator*, 217. Also see Peale's own description in the guidebook to his Philadelphia museum. *Peale's Museum*, 6–7. Another popular collection of corals was held by Boston Aquarial Gardens. Many collections featured fossilized corals (rather than specimens from living reefs), such as New York's natural history museum.

85. Coral reef specimens were a popular subject for stereoscopic photography, which had mass commercial appeal in the middle decades of the nineteenth century. Davidson, *Photography*, 2. As Davidson explains this format was particularly suited to "close-up views" of more "intricate" natural objects, such as shells and coral in particular. Davidson, *Photography*, 53, 54. Another American photographer who created coral stereographs is Joseph L. Bates. I will return to Holmes's coral stereograph in chapter 3.

86. Gilman, *Drops*, 73–74.

87. Gilman, 76, 77.

88. Darwin, *Structure*, 1, 64.

89. Dawson, *Undercurrents*, 57.

90. In one of many examples, Darwin writes that he "enquired from several native chiefs" about possible changes in the land. Darwin, *Structure and Distribution*, 138. For a discussion of Darwin's relation to Indigenous persons and knowledge in his later writings, see Beer, *Open Fields*, 55–70.

91. Dana, *Synopsis*, iv. Dana also clarifies that the specimen of *Corallium secundum* in particular "was procured at the Sandwich Islands; but it may have been brought by shipping from some other locality." Dana, *Exploring Expedition: Zoophytes*, 641. See the introduction for a discussion of the role of Black and Indigenous labor, knowledge, and skills in the practice of natural history, and in the formation of knowledge in the West more broadly.

92. Dawson, *Undercurrents*, 18.

93. Catesby, *Natural History*, 1: xlii. For more on Catesby's relation to Indigenous knowledge, skills, and labor, see Parrish, *American*, 242–46, 254–56; Delbourgo, *Collecting*, 240–44; Meyers, *Empire's Nature*, 251–61; and Iannini, *Fatal Revolutions*, 99–105.

94. Dawson, *Undercurrents*, 70. It should be noted, though, that Dawson does not mention Catesby. I draw much on Dawson's work in this portion of the chapter because it is the only scholarly work I know that sheds significant light on the historical practice of diving by Black and Indigenous persons in the African diaspora. Though Dawson mentions coral only in passing on about five occasions, on one of those occasions he mentions that enslaved persons did dive for coral—"seastone" as it was sometimes called—as a building material for planter's houses. Dawson, *Undercurrents*, 135. This practice is described by Mary Prince in her *History* (1831), in which Prince relates that at Grand Turk in the West Indies her enslavers sent a group "to break up coral out of the sea." Prince, *History*, 11.

95. Dawson mentions Hughes's knowledge of local divers in Barbados, which Hughes discusses in the natural history, a text owned by George Washington and published shortly before Washington's trip to Barbados in 1751. Dawson, *Undercurrents*, 70. Naturalist Hans Sloane likely purchased his coral specimens "from Indian and African divers" in the Caribbean, according to James Delbourgo. See Delbourgo, *Collecting*, 118.

96. Moore, *Journal*, 76, 77.

97. Dawson, 70.

98. Honeyman, "Sea-Gardens," 341. Honeyman continues, explaining of the sea-glass that "by its use the rippling of the water is overcome and one is enabled to look steadily downward, apparently to the sea-floor itself, and to see every smallest object quite as plainly as we see the things about us in the upper air. Our first glimpse gave us the impression that the glass was possessed of magical power and we gazed spellbound into the new world of grace and beauty revealed to us by its small transparent square." Honeyman, "Sea-Gardens," 341.

99. Dawson reproduces and discusses this image. Dawson, *Undercurrents*, 71. I believe that the specimens in Stark's image are coral (rather than sponge, which requires tongs to detach, according to Stark) because the image is placed within Stark's discussion of coral reefs, and just before the image Stark briefly describes the "sea gardens" of Nassau harbor as "a fine example of growing corals, where, with a water glass all the wonders of the deep can be seen as plainly as if no water intervened." Stark, *History*, 229. Precisely when Black divers in particular began to participate in these excursions is unclear to me, though certainly the practice was widespread by the late eighteenth century when, Dawson relates, tourists expected to be entertained by conch and sponge divers. Dawson, *Undercurrents*, 70.

100. I suspect that Stark's engraving is not original to this text and that perhaps Homer saw it.

101. Cooper, *Winslow Homer*, 133. For a discussion of Homer's longstanding interests in presenting Black figures with "dignity, centrality, and force," see Wood and Dalton, *Winslow Homer's Images*.

102. Cooper discusses Homer's cropping and centering. Cooper, *Winslow Homer Watercolors*, 139. Cooper also reproduces both watercolors and discusses Homer's trip to the Bahamas and Cuba in 1884–85. Cooper, 130–49.

103. For studies of US writings in the "local color" mode that portray specific regions as delightfully backward, see Brodhead, *Cultures*, 115–41. On literary representations of the US South in particular as the country's charming, backward, and rather doomed periphery, particularly during and after Reconstruction, see Greeson, *Our South*, 227–72; and McIntyre, "Promoting the Gothic South." Other images by Homer do appear in Church's essay.

104. Church, "Midwinter Resort," 499.

105. Church, 499, 506.

CHAPTER 2: "LABORS OF THE CORAL"

1. Fuller, *Letters*, 174.
2. Marx, *Collected Works*, 339.
3. Orwell, *Nineteen Eighty-Four*, 81.
4. Orwell, 122.
5. Orwell, 183–84.
6. Elleray, *Victorian Coral Islands*, 36–47. Elias, *Coral Empire*, 18, 17. Elias also mentions in passing that, though some writers mobilized coral's unpredictable growth to warn against "the potential chaos and randomness of expansionism," it was only with the emergence of twentieth-century concerns about "worker exploitation" that writers began to use the reef analogy to alert people to the plight of the laboring classes. Elias, *Coral Empire*, 19.
7. Elleray, *Victorian Coral Islands*, 39–40.
8. Elleray, 40.
9. See the introduction to this book.
10. Rockman, *Scraping By*, 8, 4.
11. Baucom, *Specters of the Atlantic*, 7. My work is indebted to Baucom's demonstration of the silenced centrality of transatlantic slavery to the "history of modern capital, ethics, and time consciousness." Baucom, 31. Baucom finds "the specter of slavery" in the late eighteenth-century rise of "a modern system of finance capital capable of converting anything it touches into a monetary equivalent." Baucom, 7.
12. As Nikhil Pal Singh explains, the "differential ethical and material valuation of human subjects" that arose to rationalize chattel slavery also enabled the dramatic rise of capitalism as a system of

appropriation and dispossession that concealed its origins in slavery. Singh, *Race*, 77. The logic of these material valuations and devaluations was sometimes as close as a coral specimen or a bracelet bead.

13. See this book's introduction for a discussion of the distinction between slavery and wage-labor, along with the reasons for discussing these two forms of labor alongside one another.

14. Barrett, *Conditions*, 186.

15. Agassiz, "The World," 217. Agassiz had earlier warned children not to speak of coral insects in her very widely read and reprinted *First Lesson in Natural History*, 25–26. For additional information on Elizabeth Agassiz and corals see Baym, *American Women*, 98–102; and Irmscher, *Louis Agassiz*, 287, 309. Adkison provides an insightful discussion of Agassiz's writing about marine life more generally. Adkison, "Elizabeth Agassiz," 9–10.

16. Tenney, *Pictures*, 98.

17. "Coral," 575.

18. Holder, "Un-Natural History," 844.

19. Quin, *Building*, 13; Hickson, *Introduction*, 10. I have included only a sampling of US literature's many warnings against coral insects. The warning was not, however, confined to the US: the most histrionic example I know is that of British socialist and biologist Edward Bibbens Aveling in *The Student's Darwin* (1881), an atheistic interpretation of Darwinian evolution for a popular audience. Aveling's plea runs, in part, as follows: "True is it that generations of men have spoken of coral insects, have had faith in coral insects: true is it that the phrase has been good enough for our fore-fathers . . . true is it that, in taking away that phrase, we are loosening humanity from its old moorings. . . . But the truth is great, and shall prevail in this as in larger matters. Let us then away with the old stereotyped phrase that has concealed the ignorance of the past. . . . There is no coral insect." Aveling, *Student's Darwin*, 51.

20. Holder, "Un-Natural History," 844.

21. Dana, *Exploring Expedition: Geology*, 80–81.

22. Dana, 80, 81.

23. "Wonders of the Deep: Coral and the Coral-Makers," 372. Judd (1890) blames writers of "prose and verse." Judd, *Structure and Distribution*, 3. Poets and philosophers are blamed by the author of "Wonders of the Deep" (1856), who laments of the coral insect that "poets have sung its praise and philosophers have speculated on its marvellous doings." "Wonders of the Deep," 371. Holder (1895) regrets that "when the public speaker or average clergyman uses the wonders of nature to illustrate his sermon, and refers to the coral, it is almost invariably to the 'coral insect'." Holder, "Un-Natural History," 844.

24. Holder (1895) deems it one of many "errors"; Agassiz (1869) explains the "coral insect" as "a mistake." Holder, 844; and Agassiz, "World We Live On," 217.

25. See Introduction for a discussion of the popular uptake of science in the US. Dobbs studies the popularity and intensity of debates about coral and reef formation in particular during the second half of the nineteenth century. Dobbs, *Reef Madness*. See also Sponsel, *Darwin's*, 19–24. The US popularity of Darwin on coral is a topic of the next chapter.

26. On the popularity of Agassiz's lectures, see Dobbs, *Reef Madness* 26; and Irmscher, *Louis Agassiz*, 87–94.

27. Dana, *Exploring Expedition: Geology*, 81. Dana's withering assessment of Montgomery's scientific acumen appears here in tiny font in a footnote to Dana's discussion of improper terms for coral. Dana repeats that note verbatim a few years later, in 1853, in *On Coral Reefs and Islands*. Dana, 48. And when Dana expanded and adapted this work for a wider audience in 1872, retitling it *Corals and Coral Islands*, he also expanded his critique of Montgomery and moved it out of the endnotes and into the main body of the text, where it remained in later editions of 1874 and 1890, as if to turn up the volume. Dana, *Corals and Coral Islands*, 19.

28. "Montgomery's Pelican Island," 98–99. This essay is a response to an earlier essay in *Norton's* by Dana, "From Professor Dana" (May 1853).

29. Today in the US Montgomery's epic is almost entirely forgotten, along with his other writings. Yet it is hard to overstate the poem's popularity in the US before 1900. For example, *Norton's* editor agreed that the "erroneous impressions" communicated by *Pelican Island* should indeed be "counteracted,"

though "to give up one's pleasure in this beautiful poem may be like losing a hand or an eye." "The 'Science' of Poetry," 82. Many excerpts of Montgomery's poem appear in US editions of *Poetical Works of James Montgomery*. Moreover, the epic received glowing early reviews in major American periodicals, and it went through many reprintings on both sides of the Atlantic well into the 1850s. Thereafter it was excerpted, alluded to, or described in detail throughout the century in countless texts across multiple genres.

30. Montgomery, *Pelican Island*, 18, 19, 21.

31. "Literary: The Pelican Island," 67.

32. "Literature, &C: Montgomery's New Poem," 105.

33. "James Montgomery," 463; Smith, *New Geography*, 83. See also the periodical essay "A Paper on Corals" (H.T.C., 1848), which uses lines from Montgomery's poem as its epigraph, then proceeds to describe the works of the "coral insect" (as though that's what Montgomery's lines also describe). "A Paper on Corals," 333. Tocque describes "the labors of the coral insects," citing lines from Canto II as an illustration of these labors. Tocque, *Mighty Deep*, 45, 46. Gilman describes "coral insects," then uses passages from *Pelican Island* as an example of their work. Gilman, *Drops*, 95, 85.

34. Bryant, *Library*, 475–76.

35. Bryant, 475.

36. "A Paper on Corals," 333.

37. Sigourney's poem was widely reprinted in part or in full in US newspapers, periodicals, poetry anthologies, natural histories, schoolbooks, and other texts at least through the 1870s. There is no full publication history of the poem, but it would include Samuel Kettel's *Specimens of American Poetry* (1829; 2:208–09); Boston's *The Conduct* (1826); Baltimore's *The Patriot* (1826); Albany's *The Escritoir* (1826); *Genius of Universal Emancipation* (1826); *New York Mirror* (1826); *Philadelphia Album* (1828); *Rural Repository* (1831); *Nantucket Inquirer* (1834); *Illustrated Poems* (1849); *Gems of Poetry* (1850); *The Mighty Deep* (1852); *Rhetorical Manual* (1853); *Arthur's Home Magazine* (1868); *A Library of Poetry and Song* (1870); and Longfellow's poetry anthology (1876). The poem has been more recently published in an anthology edited by Harold Bloom and Jesse Zuba. *Religious Poems*, 52–53. Sigourney's work has begun to receive new attention from a number of scholars, including Nina Baym (*American Women*), Mary Loeffelholz (*School to Salon*, 32–64), Wendy Dasler Johnson (*Antebellum American*, 76–107), and Jennifer Putzi, who makes the case for Sigourney's status as "indispensable" to antebellum readers' "engagement with poetry." Putzi, *Fair Copy*, 25.

38. Sigourney, *Poems*, 16–17.

39. Barrett, *Conditions*, 293.

40. Barrett, 107.

41. Baucom, *Specters*, 7. Barrett, *Conditions*, 186.

42. That lesson is glancingly present in "The Coral Insect" and *Pelican Island*: the former states that reefs "mock" the "wisdom of man," while the latter asks, "Compared with this amazing edifice, / Raised by the weakest creatures in existence, / What are the works of intellectual man?" and then presents "works" that pale in comparison, including the Tower of Babel, Babylon, and the pyramids. Montgomery, *Pelican Island*, 19–21.

43. In this way the texts carry the "cultural fable" of coral that Elleray discusses. Elleray, *Victorian*, 36–47.

44. "The Coral Insect," *Boston Pearl*, 99.

45. "The Coral Insect," 99.

46. "The Coral Insect," 99.

47. "The Coral Insect," 99.

48. "The Coral," *Harbinger*, 35, ll. 29–32.

49. "The Coral," 35, ll. 29–32.

50. Thompson, "The Coral Worm," 5.

51. Thompson, 5.

52. Poe, "Instinct vs Reason," 370; and Melville, *Moby-Dick*, 387. Melville's "orbs" may mean here land masses, celestial bodies, or the earth itself, based on possible usages listed in the *Oxford English Dictionary* (*OED*). *OED*, s.v. "orb (*n.*)," 1, 2a, 2b. (All references to the OED are to the online edition.)

53. According to a short and widely reprinted sketch announcing the birth of "A Sixth Continent" (1826) in the Pacific, "the diminutive builder . . . is at work" and will possibly "render our settlements in New South Wales of still more eminent importance." "Sixth Continent," 5. The idea that coral insects create continents comes up repeatedly in early US literature; as the author of a short sketch titled "The Coral Insect" (1835) puts it, if "*Islands* have been thus formed, why may not also the *continents?*—and why should not this hypothesis lead to a new theory of the creation?" "Original Pencil Sketches," 305, italics in original. An essay called "Coral and Coral Islands" (1835) in Philadelphia's *The Friend* speculates that, "of all the agents that are gradually and silently effecting changes upon the surface of our planet, the most universal and important is the insect which produces coral." "Coral and Coral Islands," 281.

54. Johnston, "Circulation of Matter," 553.

55. Best, *Works of Creation*, 81. On Evangelical natural history see Sivasundaram, *Nature*.

56. Best, *Works of Creation*, 84.

57. Gilman, *Drops of Water*, 84.

58. Gilman, 84.

59. Andrews, "Sea-Life," 333. Andrews's story was also reprinted in an independent volume titled *The Stories Mother Nature Told Her Children* (1889).

60. Andrews, "Sea-Life," 333.

61. Andrews, 335, 337.

62. Andrews, 333.

63. Andrews, 334.

64. Andrews, 334. Italics in original.

65. Andrews, 335, 336, 338.

66. Andrews, 333.

67. Alcott, "Whale's Story," 141.

68. U. F., "A Talk about Coral," *Woodworth's*, 14. "A Talk About Coral" by one "U.F." appeared both in *Woodworth's Youth's Cabinet* and in *Forrester's Boys and Girls Magazine*, though the original and uncredited source of much of this description is a chapter on coral polyps in English author Anne Wright's *The Observing Eye; Or, Letters to Children on the Lowest Divisions of Animal Life* (1851), a highly acclaimed, popular natural history published in London. Wright, *Observing Eye*, 40–50.

69. U. F., "A Talk about Coral," 14–15.

70. U. F., 15.

71. "A Day," *Friend*, 377. The essay probably originally appeared in the *London Christian Spectator*, according to the credit line in the two US periodicals that print it in 1864.

72. "A Day," *Friend*, 377.

73. "A Day," 377.

74. "A Day," 377.

75. "A Day," 377.

76. "Greatness of Little Things," 159.

77. "Greatness of Little Things," 159.

78. "Greatness of Little Things," 159.

79. Maury, *Physical Geography*, 153; *OED*, s.v. "hod (*n.*)," compound C2: "hod-carrier (*n.*)."

80. Maury, 153, 165.

81. Brown, *Speech*, 26.

82. Brown, 26.

83. Brown, 26. For a useful discussion of the central arguments of Brown's speech, see Peterson, *Freedom and Franchise*, 69–74.

84. Brown, *Speech*, 25. Brown elsewhere declares in his speech, "What I wish to contend for here is not so much the mere emancipation of the black race, as it is what I will define to be THE EMANCIPATION OF THE WHITE RACE. I seek to emancipate the white man from the yoke of competition with the negro. I aim to relieve the free man from conflict with the slave." Brown, *Speech*, 25.

85. "'For the slave is the coral-insect of the South,'" said the voice within; "'insignificant in himself, he rears a giant structure—which will yet cause the wreck of the ship of state, should its keel grate too closely on that adamantine wall.'" Warfield, *Miriam Montfort*, 266.

86. *OED*, s.v. "form (*v.*)," 1, 1a, 6a.

87. "Greatness of Little Things," 159.

88. Lustig, "Ants," 284. Daston describes this insect tradition, not mentioning the "coral insect" however. *Moral Authority*, 9–10, 100–126

## INTERCHAPTER: THE KORL WOMAN

1. Davis, "Iron-Mills," 74.

2. Davis, 42, 40.

3. Davis, 48.

4. Davis, 42, 55, 74. On the korl woman's centrality and provocation of the story's most pressing questions, see Pfaelzer, 41–51.

5. Davis, "Iron-Mills," 47, 55.

6. Davis, 48; and Tichi, 18.

7. Davis, "Iron-Mills," 48, 52.

8. Hood, "Framing," 76.

9. Pfaelzer, *Parlor Radical*, 42–43. Bill Brown offers an extensive discussion of the "korl woman," and of the sense of "timeliness" and urgency that Davis means for this figure to evoke, and that she underscores through the particular material from which Hugh carves it. Brown, "Origin," 787.

10. Davis, "Iron-Mills," 40–41.

11. "What hope of answer or redress?" asks the epigraph, which Davis adapts from Tennyson's *In Memoriam A.H.H.* (1850). Davis, 39.

12. Davis, 74.

13. For information on this periodical in particular see Ranta, *Women and Children*, 52–53. As Ranta explains, "Like the *Lowell Offering*, *The Operatives' Magazine* published more controversial material than it is usually given credit for, including pieces exploring problems of class bias and illness and death caused by factory work. The editors' intentions were to exhibit the abilities and talents of the operatives and to help them develop their talents." Ranta, 52. For more on the political importance of workingwomen's periodical literature see Merish, *Archives of Labor*, 51–60.

14. Anna, "The Coral Insect," 191.

15. *OED*, s.v. "wreathe (*v.*)," 8a.

## CHAPTER 3: FATHOMLESS FORMS OF LIFE

1. Peyssonnel, "An Account," 454. Peyssonnel's treatise was never published in full, but selections were translated into English by William Watson and read before the Royal Society in 1752. It is Watson who translates Peyssonnel's "urtica, purpura, or polype" as "coral-insect." Peyssonnel, "An Account," 454.

2. For a discussion of Peyssonnel's contributions to Enlightenment coral science, and a useful discussion of the larger debate about coral taxonomy, see Gibson, *Animal, Vegetable, Mineral?* 120–28. Two other sources from which I have drawn information on Peyssonnel's biography and discoveries are Bowen, *Coral Reef Era*, 24–26; and Vandersmissen, "Experiments."

3. Historians have sometimes deemed the *eighteenth* century the "age of classification" because Enlightenment naturalists such as Linnaeus famously described and sorted nature based on features such as color, shape, size, and texture. Yet it was only during the nineteenth century that taxonomy became fully determinative, Robyn Wiegman explains, drawing upon the work of Stepan (1982), as the goal of natural history shifted from describing a more fluid and mutable natural world to determining its fixed, biological essences. Wiegman, *American Anatomies*, 28–30. During this latter

period, for example, the Enlightenment's "Great Chain of Being" began to reemerge as a schema for identifying distinct racial "types" and grading them in descending order from the white European male down. Wiegman, 29. On this transformation toward biological determinism, and away from an earlier nominalism, arbitrariness, and fluidity, also see Dayan, *The Law*, 13–24; and Guillaumin, *Racism, Sexism*. Guillaumin observes that Linnaeus "would probably have been very surprised if one had connected him to some endo-determinism." Guillaumin, 151. Driving the shift toward the rise of nineteenth-century biological determinism in the US in particular was slavery. Ralph Bauer and José Antonio Mazzotti demonstrate that the pressure to sustain and expand an economy built on slavery rendered "race" primarily a matter of skin color and other ostensibly visible features, rather than a multiply informed and flexible indicator of human variety that changed in response to environmental influences. Bauer and Mazzotti, *Creole Subjects*, 1–58. See also Jordan, *White over Black*, 482–511; Fredrickson, *Black Image*, 71–96; and Dain, *Hideous Monster*, 197–263. Roxann Wheeler offers a particularly nuanced literary analysis of the complex and contested relations between economic conditions and racial ideology across the long eighteenth century, focusing particularly on Great Britain. Wheeler, *Complexion of Race*, 2–48.

4. Here is the full quote from Emerson: "I hear the chuckle of the phrenologists. Theoretic kidnappers and slave-drivers, they esteem each man the victim of another, who winds him round his finger by knowing the law of his being, and by such cheap sign-boards as the color of his beard, or the slope of his occiput, reads the inventory of his fortunes and character. The grossest ignorance does not disgust like this impudent knowingness." Emerson, *Letters*, 475. I am especially inspired by Dominic Mastroianni's insightful interpretation of this passage as an expression of epistemological skepticism and by Mastroianni's larger analysis of the epistemological optimism—defined briefly as a belief that everything that matters is knowable—shaping much antebellum literature and culture. Mastroianni, *Politics and Skepticism*, 27–28; 1–7. Infamous examples of the racist application of phrenology may be seen in some of the period's most influential treatises on race, such as Samuel George Morton's *Crania Americana* (1839) and Josiah Nott and George R. Gliddon's *Types of Mankind* (1854), both of which Dain discusses at length. *Hideous Monster*, 197–226.

5. Racist ideas were, after all, never the cause of slavery; for a clear and accessible discussion of this fact, see Kendi, *Stamped from the Beginning*, 9–11.

6. Resisting that impulse is an increasingly crucial cognitive practice in our own moment of racial violence. As Claudia Rankine explains, a "rerouting of interior belief" must accompany larger-scale political change. Rankine, "Condition of Black Life." Rankine's *Citizen* (2014) models and prompts that rerouting: according to Gosetti-Ferencei, *Citizen* encourages readers to resist the "habitual categorization that is characteristic of racial prejudice itself." Gosetti-Ferencei, "On Literary Understanding," 83. Coral once occasioned just such resistance, which is one of many reasons to learn to experience coral as early Americans did.

7. See chapter 1 of this book for a discussion of coral fishers and the global trade of Mediterranean red coral.

8. Bowen comments on the "floral character" of these engravings. Bowen, *Coral Reef Era*, 23.

9. Peyssonnel, "An Account," 454. And according to Watson's translation Peyssonnel further explains "that the animal, when it *wanted* to come forth from its niche" pressed outward and appeared like a flower, an appearance that had encouraged Peyssonnel's predecessors to mistake coral for a plant. Peyssonnel, 455.

10. Reaumur is cited in Vandersmissen, "Experiments," 64. Voltaire, *Singularités*, 8. Voltaire insisted that coral is a plant because it looks like one, long after Linnaeus had been convinced otherwise. Voltaire, 9.

11. As Vandersmissen puts it, among the French, Peyssonnel "sank into oblivion." Vandersmissen, "Experiments," 65.

12. For a discussion of Trembley's discovery of the polyp, along with some of the new taxonomic, biological, epistemological, and spiritual questions that this discovery raised, see Dawson, *Nature's Enigma*, 85–136; Roger, 316–18; and Lenhoff and Lenhoff, *Hydra*. Danielle Coriale reveals that the polyp's status as a creature that "defied natural laws, disrupted epistemological assumptions, [and]

resisted classification" inspired a number of Victorian writers "to fabricate robust naturalist fantasies." Coriale, "When Zoophytes Speak," 19–20.

13. Réaumur retracted his insult in 1742. The Royal Society of London validated Peyssonnel's findings and admitted him as a Fellow in 1752. In 1756 Peyssonnel even returned briefly to France where he was admitted to and invited to speak at the *Académie*. "One cannot address any of these productions without recalling the name and discovery of M. Peyssonel," reflects one of the entries on coral in Diderot's *Encylopedia*. Daubenton, "Coral."

14. In his own work on coral, Thomas Henry Huxley (1870) writes of Peyssonnel's retirement to the West Indies: "and finally he took himself off to Guadaloupe, and became lost to science altogether." Huxley, "On Coral," 114.

15. As Bowen explains, though Linnaeus classified some corals as a kind of animal (Anthozoa) in the tenth and definitive edition of *Systema Naturae*, he also continued to insist that many have, in Linnaeus's words, "a mere vegetable life"—even after Ellis asserted in print that corals are not the "mere vegetables" that even "very sensible and curious Naturalists" continue to mistake them for. Bowen, *Coral Reef Era*. Ellis, *An Essay*, xiv.

16. Akin, *Encyclopaedia*, 443.

17. Wakefield, *Mental Improvement*, 60, 61. This 1799 edition is the first American edition of the text.

18. Miller, *Retrospect*, I: 124. Waterman calls Miller's work an "encyclopedic investigation of European and American arts and sciences" produced collaboratively by members of New York's Friendly Club, including Charles Brockden Brown. Waterman, *Republic*, 233. Another source of the popularity of coral insects and Peyssonnel in the early US may be Erasmus Darwin's writings, which were published in US editions and very widely read by Americans. Darwin mentions coral insects on several occasions, even referring readers of *The Botanic Garden* (1791) to Peyssonnel's treatise, as translated by Watson and presented before the Royal Society, for more information on the coral insect. Darwin, *Botanic Garden*, part 1, 123.

19. Dana, *Exploring Expedition: Zoophytes*, 7.

20. Dana, 7–8.

21. Dana, 7–8.

22. Schele de Vere, *Wonders of the Deep*, 104, 105.

23. Schele de Vere, 105–6.

24. A final example that rivals Schele de Vere's in dramatic force is *Leisure Hours Among the Gems* (1884), a popular mineralogical treatise by former Union Army medical inspector Augustus Choate Hamlin. Hamlin's text bears a dedication to Peyssonnel, "who ventured to announce to the men of science of the Royal Academies of Europe in the eighteenth century that the CORAL was the product of animal life, and not of vegetable growth. In answer to his simple discovery and honest declaration, the naturalist was met with a storm of contempt and derision that eventually wrecked his happiness and his life." Hamlin, *Leisure Hours*, 5. Just as in Schele de Vere's account, here Peyssonnel suffers inordinately and beyond redress on coral's behalf, "wrecked" by his "simple discovery" and "honest declaration." When Hamlin later returns to the story of Peyssonnel in the text itself, he draws the lesson that we need not repeat the same mistake today: "When we come to review the hypotheses of science during the last century, we shall feel more inclined to be generous and flexible in our views of natural phenomena. 'There are more things in heaven and earth, Horatio, / Than are dreamt of in your philosophy'." Hamlin, 72. A tragedy of Shakespearean proportions, the story of Peyssonnel warns us, as Hamlet warns Horatio, to remember the very limits of human knowledge itself, and especially when it comes to assessing "natural phenomena."

25. Schele de Vere, *Wonders of the Deep*, 105.

26. *OED*, s.v. "social (*adj.* and *n.*), compound C2: "social status." Italics mine.

27. Hazard, "Evangelical Encounters," 203. On British evangelical interpretations of coral, see Elleray, *Victorian Coral Islands*.

28. Coxe, *Wonders*, 96. As Coxe explains in her preface, this work is indebted to Hale's *Good's Book of Nature* (1834), among other sources. Coxe, iii. I can find almost no information on Margaret Coxe

(1800–1855), author of the two-volume *Claims of the Country on American Females* (Columbus, 1842), which I discuss later in this chapter.

29. Coxe, *Wonders*, 94.
30. Coxe, 7, 94–95.
31. Coxe, 96–97.
32. Coxe, 95.
33. Coxe, 96.
34. That same lesson, in adumbrated form, appears in multiple other texts. One example is an essay called "Coral" in Boston's *Merry's Museum* (1847): beneath an engraving of a coral branch we learn that coral "was for ages thought to be a vegetable product; but about a century ago it was found to be the work of a *living animal*" called "the coral insect." "Coral," 3.
35. Coxe, *Claims*, vol. 1, 17. Additional discussion of the place of women in the "scale of being" occurs elsewhere in this text. Coxe, vol. 2, 20, 117.
36. Kidder, *Coral-Maker*, 4. This text has a complex, transatlantic publication history. In 1844 it was issued in London and New York as an independent volume, "revised by the editor, Daniel P. Kidder," according to the title page. The volume's contents were originally published as a chapter called "The Coral-Maker" in *Wonders of the Waters* (1842), by the Religious Tract Society of Great Britain. According to a review in London's *Christian Lady's Magazine*, "The Tract Society's last two little square books are on 'The Animalcule,' and 'The Coral-maker,' both most interesting: the latter especially exhibits a scene of wonder surpassing all that have preceded it in this instructive series." "Review of Books," 467. Kidder, an American Methodist Episcopal theologian, must have found the tract on coral compelling enough that it would be widely embraced and used in US Sunday Schools. For an illuminating and related discussion of this pamphlet, see Chow, *Nineteenth-Century American Literature*, 102–3.
37. Kidder, *Coral-Maker*, 3.
38. Blake, *Wonders*, 116. Blake's passage on coral appears to be adapted from an essay originally published in London's *Universal Review* (and then reprinted in US periodicals including *Arthur's Home Magazine* in 1868) under the title "Works of the Coral Insect" (1868, 302).
39. Several performance notices appeared in Boston newspapers. For crucial background information on early American singing and its popular uptake I am indebted to conversations with Laura Stokes, Music Librarian, Brown University, and email exchanges with Marian Wilson Kimber, whose *The Elocutionists: Women, Music, and the Spoken Word* informs my discussion of Hale's song and helped me track its performance history and incredibly widespread influence on other poems and songs published thereafter. First published in Hale's *School Song Book*, a volume of songs intended for "social singing," the song was soon scored by well-known Boston composer George James Webb, printed in a popular educational periodical, performed at several public concerts held by the Boston Academy of Music, published as a poem in newspapers and magazines, and reprinted in widely used song books at least four times between 1836 and 1872. Hale, *School Song Book*, 16, iii; and Webb, "The Coral Branch," 124. Furthermore, the influence of this song is discernible in the language of several other poems and songs published in the US well into the 1860s. See, for example, Slade, "The Coral Insect; or Perseverance," 46. To listen to a contemporary adaptation and recording of Hale's song by Ian Crockett (lead vocal) and Gene Navakas (guitar and additional vocals), visit "The Coral Branch." Michele Navakas (personal website): https://michelenavakas.com/thecoralbranch.
40. Hale, *School Song Book*, 16.
41. Peyssonnel, "An Account," 7, 10.
42. Hale, "Coral Branch," 10, 11, 15, 16, 18, 19.
43. Here Hale also suggests that reefs manifest a form of labor that extends the self—but in a way that does not exclude, but rather sustains, others. I will return to the concept of expansion by way of sustaining in chapter 4.
44. Finley, *Lady of Godey's*, 97.
45. *Good's Book* was first published at London in 1826. This abridged version, first published at Boston in 1834, went through multiple editions. As M. Sarah Smedman observes, "although [Hale's] name

does not appear on the juvenile version of *Good's Book of Nature* (1834), the adaptation has been attributed to her. She had recently been introduced to the textbook field through her cooperation with Lowell Mason, and certain stylistic qualities in this edition of John Mason Good's book are characteristic of her." Smedman, "Sarah Josepha (Buell) Hale." Isabelle Webb Entrikin earlier attributes the edition to Hale, and for a nineteenth-century attribution see Wheeler, who cites *Allibone's Dictionary of Authors* as listing the abridgement as "edited by Mrs. Hale." Entrikin, *Sarah Josepha Hale*, 137; and Wheeler, *History of Newport*, 128. I quote here from the eleventh edition of the abridgement (1843).

46. Hale, "General Resemblance," 40.

47. Hale, 40.

48. Hale, 40.

49. Guillaumin, *Racism*, 71.

50. Darwin, *Structure and Distribution*, 1, 12, 14, 64.

51. During this period the study of coral formations was valued mainly as a way to understand the history of the motion of earth's crust. "Barrier reefs and atolls were markers of the former presence of land" whereas fringing reefs indicated that the crust was either relatively stable there, or was rising. Herbert, *Charles Darwin*, 234. Yet, as Janet Browne observes, Darwin's reef geology was always also biological: according to Darwin, "biological and geological forces acted in concert to bring about what he called one of the most remarkable alterations in the natural world." Browne, *Voyaging*, 319. The location, rate, and results of those alterations depended much on coral biology, according to Darwin: as Rudwick explains, Darwin posited that "provided a continent or an ocean floor sank slowly enough, the coral organisms could build the solid [reef] structure upwards, keeping pace with the sinking foundations and maintaining themselves in the shallow water they were known to require." Rudwick, *Worlds before Adam*, 491. Darwin's theory of coral formation has since been variously tested and revised in light of new knowledge about other factors that influence the growth of coral, including changes in sea level.

52. Darwin, *Structure and Distribution*, 1, 64. See, for example, a discussion of Darwin's treatise in an 1846 essay in the *Christian Register*, "The Coral Insect," 156; the reference to Darwin's polyps as "coral insects" in the 1858 essay, "Coral Reefs" in *Godey's Lady's Book*; and "The Busy Coral Insect," an 1889 *New York Times* review of a later edition of Darwin's coral treatise.

53. Darwin, *Structure and Distribution*, 1, 52, 63.

54. Darwin, 24.

55. Darwin, 25, 36.

56. A major influence on Darwin's thinking about nature more broadly, and likely coral in particular, is Alexander von Humboldt, who describes coral islands as continually emerging from the work of the "harmonious lithophytes" (polyps, presumably), seeds, winds, and waves. Humboldt, *Ansichten*, 181. For Humboldt's conception of nature, see Walls, *Passage to Cosmos*, who discusses the relation between that conception and Darwin's. Walls, 236–38. For more on the influence of Humboldt's work on Darwin, see Browne, *Voyaging*, 133–36.

57. Grosz, *Time Travels*, 15, 25, 41.

58. Grosz, 17–18.

59. Grosz, 36.

60. Hale, *Godey's*, 240; "The Coral Insect," *Christian Register*, 156. For just one example of a sermon that imports Darwin's ideas on coral see Thompson, *Moral Unity*, 36–37, 60–71. Chow discusses Lyell's US lectures, which spread Darwin's ideas about coral. Chow, *Nineteenth-Century American*, 105–6.

61. Darwin, *Structure*, 47.

62. Montgomery begins his poem's preface with the claim that "the subject of 'The Pelican Island' was suggested by a passage in Captain Flinders's *Voyage to Terra Australis*." Montgomery, *Pelican Island*, v. A very short sampling of US texts that import Flinders's island would include Blake, *Parlor Book*, 217–18; Goodrich, *Peter Parley's Wonders*, 137–38; "Formation of New Islands," 359; and "A Conversation About Islands," 74. Flinders's US popularity may be seen as part of the larger trend of US interest in writing about the Pacific that Paul Giles analyzes in *Antipodean America*.

63. Flinders, *Voyage*, 2:115. Entry for 1802, October, Saturday 30.
64. Flinders, 115.
65. Flinders, 116.
66. Montgomery, *Pelican Island*, 22.
67. "Coral Islands" (1823), 400.
68. Goodrich, *Peter Parley's Wonders*, 138.
69. "The Science of Geology," 409.
70. Grosz, *Time Travels*, 15. Grosz acknowledges that the steep challenges to a feminist uptake of Darwin would include, most notably, that his theory of evolution has been interpreted as one "of 'winners and losers,' of those who dominate and those who have succumbed to domination or extinction, a theory that, on the face of it, seems to provide a perfect justification for the relations of phallocentric and racist domination that constituted Eurocentric, patriarchal culture in his time as much as in ours." Grosz, 16. While recognizing the sexism and bias of Darwin's account, however, Grosz asks a question that has guided my interpretation of Darwin's coral treatise: "Without necessarily minimizing these investments in male and white privilege, do these discourses provide theoretical models, methods, questions, frameworks, or insights that nevertheless, in spite of their recognizable limitations, could be of some use in understanding and transforming the prevailing structures of (patriarchal) power and in refining and complexifying feminist analyses of and responses to these structures?" Grosz, 17. Another problem, of course, is that Darwin offers a largely biological account of life, and during the nineteenth century biologism frequently sponsored colonial violence. Yet, as Cristin Ellis reminds us, "biologism does not inherently fund discrimination," which is why certain nineteenth-century antislavery writers grounded their challenges to discrimination in a biological, materialist account of persons. Ellis, *Antebellum Posthuman*, 5–6. Stacy Alaimo's (2000) interpretation of Darwin's later writing (published long after his coral treatise) is useful in this regard: Alaimo observes that when "reading Darwin, we imagine our own body, no longer an entity unto itself, as a site of metamorphosis, still bearing traces of its forebears, the 'hairy, tailed quadruped,' the 'amphibian-like creature,' the 'fish-like animal,' and finally, the 'aquatic animal provided with branchiae, with the two sexes united in the same individual . . .'." Alaimo, *Undomesticated Ground*, 41. That materially grounded way of thinking about bodies could possibly abet a feminist politics.

## INTERCHAPTER: "I COME FROM CORAL REEFS"

1. Nichols, *Long Memoried Woman*, 8–9.
2. Nichols, 8.
3. Neumann and Rupp, "Sea Passages," 478. See also Narain, who describes Nichols's "interest in the body as a site of densely coded meanings and inscriptions." Narain, *Contemporary Caribbean*, 183.
4. Bennett, *Being Property Once*, 171; 170.
5. Bennett, 189.

## CHAPTER 4: CORAL COLLECTIVES

1. Latimer, "Coral Creations," 117.
2. Latimer, 117.
3. During the nineteenth century, as in our own time, the degree to which the pursuit of self-interest would produce common good—and even exactly what "self-interest" constituted—was a topic of much deliberation among politicians and the broader public, as a number of scholars have shown.
4. As Jeanne Boydston argues, "women's unpaid domestic labor" was "a central force in the emergence of an industrialized society in the northeastern United States," yet "the society these women [lived]

in—and, in many respects, the women themselves—had come to doubt, even to deny, the economic value of their labors," in comparison to "the wages of the husband." Boydston, *Home and Work*, xi, xii.

5. Boydston, 201. On the erasure of labor from social and economic roles traditionally assigned to women, also see Boydston, *Home and Work*; and Doyle, *Maternal Bodies*.

6. As literary scholar Dana Nelson crucially reminds us, "varieties of exclusion" frequently haunted the collectives that nineteenth-century Americans imagined or enacted as alternatives to liberal individualism. Nelson, *Commons*, 10.

7. A long-standing critique of white feminism, of course, has been its failure to recognize that the experiences of Black women are fundamentally distinct. For more on this topic, see Davis, *Women, Race & Class*.

8. Coral thus helped these writers and activists theorize and promote some of the "practices of citizenship" that Spires identifies as central to nineteenth-century Black citizenship, an alternative to a white version of citizenship from which Black people were excluded. Spires, *Practice*, 1–3.

9. As Davis reminds us, "special abuses" were inflicted on enslaved women to "[facilitate] the ruthless economic exploitation of their labor," and these experiences set them apart from white women, even if both groups were subjected to certain shared preconditions of capitalism. Davis, *Women, Race & Class*, 7. Following these insights, I have taken care to distinguish between the distinct *experiences* of different groups of marginalized persons, and some of the shared *conditions* that marginalized them: it is not possible to compare experiences; it is imperative to compare conditions.

10. Barrett, *Conditions*, 100.

11. In this particular way, we might see the white women in this chapter within a longer tradition of white feminism, one goal of which, Haraway tells us, is to create "a world less riddled by the dominations of race, colonialism, class, gender, and sexuality." Haraway, *Simians*, 2.

12. "A Branch of Coral," 415. The essay continues: "Formerly, the coral branch was regarded as one animal—an individual . . . But no zoologist now is unaware of the fact that each polype on the branch is a distinct individual, in spite of its connections with the rest."

13. Beecher, "Christianity," 312.

14. Beecher, "Sermon," 2.

15. But Beecher is clear that the only way to *spiritually* sustain the self in the long run is to pursue the well-being of others, which is not what is happening in his reef; coral for Beecher is not helpful in telling us what we ought to do or be, rather, it tells us who we already are—naturally selfish people, motivated by the pursuit of individual gain, which by a happy accident enriches all.

16. Spofford, *Happiness*, 65. Spofford did elsewhere explicitly express rather "troubling race politics," observes Logan. Logan, "Race, Romanticism," 36.

17. Spofford, *Happiness*, 65.

18. Spofford, 65–66.

19. Spofford, 66, 68.

20. Spofford, 68.

21. Spofford seems to suggest that it really all comes down to one's perspective, a premise that is central to her book: as she explains in the preface, "Happiness is equally attainable to the poor and the rich, the youth and the veteran." Spofford, 3. But apparently happiness *also* relies on maintaining boundaries between those within and outside of the US, as the allusion to American Indians indicates, for the wealthy woman's security depends on troops protecting "the frontier haunted by the tomahawk." Spofford, 66.

22. Mann, "Coral," 575. The article is signed "M.," but it seems quite likely that the author is Mary Tyler Peabody Mann, a significant contributor to *Arthur's Home Magazine* and the wife of Horace Mann, who also wrote about coral in similar terms. Mann, *Lectures*, 285.

23. Beecher, "Unity of Man," 207.

24. Perhaps it is not surprising that such a political vision would emerge from accounts of coral islands in Flinders's *Terra Australis*, a work of British colonial exploration and enterprise. Flinders, 2: 115–16.

Victorian writers frequently evoked reefs as metaphors for the robustness and all-encompassing beneficence of nineteenth-century British imperial expansion, as Elias observes. Elias, *Coral Empire*, 17–19.

25. Latimer, "Coral Creations," 117. All citations from this source come from page 117.

26. Some of this language ("rooted"/"shoot out") comes from an earlier passage on coral in this same essay. Latimer, 117.

27. Harris, "Barbara E. Pope," 285. Harris asserts that *Waverley Magazine* especially attracted many African American women authors who might find themselves "excluded from other venues for personal or professional reasons." Harris, 286. Harris cites the words of *Waverley*'s editor in 1853, Moses A. Dow, who told readers that the magazine's "impartial and liberal policy, combined with its extensive circulation, furnished an avenue for the promotion and encouragement of genius before unknown." Harris, 285.

28. All quotations are from Latimer, "Coral Creations," 117.

29. Darwin, *Structure*, 63.

30. Latimer, "Coral Creations," 117.

31. As an active member of the Association for the Advancement of Women, one of the few female speakers at the Concord Summer School of Philosophy (1879–88), and thereafter the first US woman to hold the position of university lecturer, Mitchell wrote largely to and for other women. Rogers describes Mitchell as one of the period's many "paraprofessional philosophers," women who "rarely worked within the halls of the academy. Instead, they were teachers, public lecturers, social critics, and/or political analysts whose audience was almost exclusively female." Rogers, *America's First Women*, 97. As Rogers further observes, in her writings Mitchell "maintains that an isolated individual existing apart from the social order is an impossibility, 'an absolute non-entity.' In fact, social institutions play a critical role in individual self-determination." Rogers, 109. Elsewhere Mitchell writes, "I must lose this single, separate self of mine in the larger self of the family, of the state, of the race in order to attain spiritual growth and development. The social institutions that surround me, instead of limiting my freedom, enable me to transcend all that is narrow and selfish, to identify myself with other human beings and make their life my own." Qtd. in Rogers, 109. On *Arthur's* as a magazine primarily for women and children see Mott, *History*, 416–18.

32. Mitchell, "Corals," 757.

33. Mitchell, 756. Mitchell sometimes calls corals "he." Mitchell, 757. Yet she portrays corals pretty consistently in terms that describe labor gendered female, as in: "There is something mysterious and suggestive in the life of this little creature" that is always "feeding" and "producing others," its small body sustaining "countless generations" so that the reef rises "layer after layer" long after its death. Mitchell, 756. Is Mitchell thinking specifically of women when she writes cryptically that this creature's life is "suggestive"?

34. Mitchell, 757. She continues: "Were their existence not limited by certain local boundaries and fixed conditions of temperature" corals would fill the entire ocean with reefs. Mitchell, 757.

35. Mitchell, 757.

36. Mitchell, 757.

37. Mitchell, 758; and Latimer, "Coral Creations," 117.

38. Stowe, "Coral Ring," 277.

39. Stowe, 275, 273, emphasis in original. The story has a solid republication history: it was republished in 1848 in *Godey's Lady's Book*, then by the Scottish Temperance League in 1853 as an independent volume. Stowe then included it in *The May Flower and Miscellaneous Writings* (1855) and in *Stories, Sketches, and Studies* (1896). There is very little extant scholarly commentary on this short story, though Carol Mattingly usefully observes of it that "Stowe indicts a social system that wastes women's lives in 'parlor ornamentation.'" Mattingly, *Water Drops*, 208.

40. Stowe, "Coral Ring," 266.

41. Stowe, 267, 268.

42. Stowe, 268.

43. Stowe, 269.

44. Stowe, 271, 272.

45. Stowe, 274.
46. Stowe, 277, emphasis added.
47. Stowe, 278, 280, emphasis added.
48. Stowe, 281.
49. Stowe, 276.
50. Latimer, "Coral Creations," 117.
51. The image appears on the second page of Darwin's coral treatise; Darwin discusses "ring-formed" reefs beginning in part 3.
52. Qtd. in Carp, *Adoption in America*, 37. For more on the Fellows sisters see Carp, *Adoption in America*, 37–46. Their Boston journal ran from 1845 to 1857.
53. Fellows, "Small Efforts," 36. All citations from this source are page 36.
54. Fellows, 36.
55. Harper, *Poems*, 44–45.
56. The poem is "addressing black children directly" as well as "their caretakers," according to Chandler, "'Ye Are Builders',", 48, 49.
57. Loeffelholz, *School to Salon*, 104, 102.
58. Loeffelholz, 100–104.
59. "Little Builders," *Juvenile Missionary Magazine*, 143–44. The piece may have been written by Helen Taylor since it appears in her *Missionary Hymns for the Use of Children* (1846). Taylor, *Missionary Hymns*, 18–19. The *Juvenile Missionary Magazine* credits *Missionary Hymns* as the poem's source.
60. One US periodical that published this poem is *Zion's Herald* (1847), and a number of different poems with this same title appeared in US periodicals during the period.
61. Elleray, *Victorian*, 44.
62. Qtd. in Foster, *Written by Herself*, 134. For more on the political goals of Harper's Reconstruction poetry, see Carla Peterson, who explains that Harper "considers one of the fundamental problems of Reconstruction—the assimilation of African Americans into the national body politic—from multiple perspectives." Peterson, *"Doers of the Word,"* 210.
63. Loeffelholz, *School to Salon*, 102.
64. Foster, *Written by Herself*, 134.
65. Qtd. in Foster, *Written by Herself*, 134.
66. Foster explains that Harper's "other theme, and the one that increasingly dominated her published writings, was that the Emancipation had opened a new era—a time for blacks, particularly black women, to "'lift up their heads and plant the roots of progress under the hearthstone.'" Foster, *Written by Herself*, 134. Loeffelholz observes that the reef metaphor "dramatizes the prehistory and the hoped-for emergence of a new kind of national body, not yet fully visible, one in which African Americans including African American women may imagine themselves participating." Loeffelholz, *School to Salon*, 101.
67. Peterson, *"Doers of the Word,"* 213. Loeffelholz explains that Harper's poem "seems clearly to have been written in the first instance for recitation to the audiences of Harper's 1864–71 lecture tours" and is specifically directed at "Harper's audience of emancipated African American children." Loeffelholz, 101–3. Chandler adds that the poem *also* initially addresses these children's "caretakers." Chandler, "'Ye Are Builders,'" 49. Yet the poem's audience widened as it appeared in print in *Poems* (1871), a collection that, Peterson explains, "[addresses] the nation itself . . . or its most committed leaders," since many of these poems "insist . . . on the post-Civil War responsibility of whites to reconstruct the nation as a home for African Americans." Peterson, *"Doers of the Word,"* 210. For an account of Harper's remarkably ambitious Reconstruction lecture tours, see Foster, *Written by Herself*, 134. Foster reproduces some of these lectures, as well as letters in which Harper discusses them.
68. Spires, *Practice*, 3.
69. Spires, 236.
70. Spires, 3.
71. Spires, 3.

72. Smith, "Civilization," 262, 263, 262. Smith's central argument is, as Britt Rusert observes, that "the essential condition of civilization, derived from the Latin *civis*, is 'co-ivis,' a 'coming together,' which requires the assemblage and commingling of diverse peoples." Rusert, *Fugitive Science*, 53. Rusert also discusses the importance and role of *Anglo-African Magazine* as a forum for "black intellectuals, activists, educators, and theologians on statistics, history, philosophy, and diverse inquiries into natural science." Rusert, 28.

73. Smith, "Civilization," 262–63.

74. Chow, *Nineteenth-Century American Literature*, 117, 98–141. Chow discusses Smith's metaphor within the context of Smith's broad intellectual, political, and spiritual engagement with the "biogeographical" dimensions of coral science as elaborated by Lyell, Dana, and Darwin (though not in the coral treatise) and finds that coral helps Smith articulate a "concept of civilization" shaped by an "archipelagic form of dispersal and assemblage." Chow, 114–15. Cherene Sherrard-Johnson also discusses Smith's coral metaphor, observing that it emphasizes the "wonder of collective labor," and that Harper, too, evokes coral "in concert with Smith's sentiments that patience, steadiness, and sustained labor is what's needed to materialize the Black freedom dreams." Sherrard-Johnson, "Sensing Black Coral," 3, 16–17.

75. Townsend, "The Policy," 324. Townsend was actively involved in abolition on both coasts and was then serving as editor of *Frederick Douglass's Paper* according to Gardner, "Townsend, Jonas Holland."

76. Townsend, "The Policy," 325, 326.

77. Townsend, 327. Smith would have agreed with this claim, as his own and earlier use of the coral insect metaphor to describe the power and value of educating Black children suggests. Smith, Letter to Gerrit Smith, 308.

78. Barnett, "Race Unity," 603–4. For more on Barnett, see Spatz, "Barnett, Ferdinand L."

79. Barnett, "Race Unity," 606–7.

80. Barnett, 603.

81. Barnett, 607.

82. I could find very little information on DeMond, for whom there is no entry in *Oxford African American Studies Center* or in other major biographical sources. DeMond's "Negro Element" is apparently his only known publication.

83. DeMond, "Negro Element," 10.

84. DeMond, 10.

85. DeMond, 12.

86. Andrews, "Sea-Life," 33.

87. See this book's Coda for further analysis of DeMond's metaphor in this regard.

INTERCHAPTER: "THE CORAL BUILDERS,"
ST. JOHN'S AME CHURCH, CLEVELAND, OHIO

1. McHenry, *Forgotten Readers*, discusses these and other shared goals of Black literary societies, though she does not discuss the Coral Builders. McHenry, 3, 19.

2. For example, in 1867 a Jacksonville society of women called the "Coral Builders" spearheaded a campaign that eventually raised enough money to rebuild their own St. John's Church after it had been destroyed in a fire, reported a New York missionary society journal in 1897. "A Word of Encouragement," 571. In 1871 the "Coral Workers" of West Philadelphia's Woodland Church "held a little fair" that raised funds for overseas missions, according to an item in *Woman's Work for Woman*, the journal of the Woman's Foreign Missionary Society of the Presbyterian Church. "For Next Christmas," 148. In 1886 Boston's *Missionary Herald* recorded donations from a Connecticut society called "The Coral Builders." "Children's 'Morning Star' Mission," 320. In 1872 New York's *Advocate and Family Guardian*, with its entirely female editorial staff, featured a (probably fictional) sketch about a women's sewing circle that called itself "The Coral Workers" and met every two weeks to sew items that supported a local home for the city's poor. "The Coral Workers," 78–79. *Life and Light*

*for Woman*, the journal of Boston's Woman's Board of Missions, featured a laudatory letter "For the Coral Workers" and reported on the contributions of numerous small "Coral Workers" or "Coral Builders" societies—from Lyme, Connecticut to Ripon, Wisconsin—that were apparently groups of children in sabbath-schools who raised funds for overseas missions. "For the Coral Workers," 154–55, 120, 160. In 1906 Louisville's *Christian Observer* described "The Coral Workers' Society" of Lynchburg, VA, noting that it is "composed of young girls" whose work sustained Lynchburg's Third Church. "Virginia," 13.

3. Robinson, "Cleveland," 595. The Church held services in members' homes in the area of the city now known as Public Square, and then in the Apollo Hall on Merwin Square, until moving to its first dedicated building on Bolivar Street in 1850. Thereafter, through a period of significant struggle, the congregation grew as the Church moved to a number of other buildings until 1908, when the congregation built a new church at East 40th Street and Central Avenue, where St. John's presently stands.

It has been difficult to find much detailed information about the Church's history before 1900. I have relied primarily on five sources of information: Tuennerman-Kaplan's discussions of St. John's within the larger context of church-related charitable activities among African American communities in Cleveland between 1880 and 1930. Tuennerman-Kaplan, *Helping Others*, 84–100; the entry on St. John's in *Case Western Reserve University's Encyclopedia of Cleveland History*. "St. John's African Methodist Episcopal (AME) Church"; Robinson, "Cleveland," 595–99; the preface to *150th Anniversary, St. John A.M.E. Church Souvenir Journal, 1830–1980*; and some brief and unpublished documents ("The History of St. John" and "Historical Articles Related to the St. John") created by the Church and kindly shared with me by church member Charles M. Armstrong in February 2022. There is one rare published early history of the church (Hicks, *St. John A.M.E.*), though based on my correspondence with an archivist at the Western Reserve Historical Society, Hicks's coverage of pre-1900 church history is quite limited.

4. Robinson, "Cleveland," 595–6.

5. Robinson, 596.

6. For the demography of the membership of Cleveland's Baptist and AME churches during the late nineteenth century see Tuennerman-Kaplan. *Helping Others*, 90.

7. *Cleveland Gazette*, October 29, 1892. The *Gazette*'s sustained interest in this particular Society appears to bear out Felecia G. Jones Ross's findings that in these years this newspaper aimed to promote racial integration and "redress" of discrimination, legal and cultural, while it "informed its readers of Black representation and achievement in a variety of areas." Ross, "Fragile Equality," 60, 62. Indeed the *Gazette*, which is searchable through *Readex African American Newspapers*, has been my primary source of information about the Society, though why announcements end in the April 6th, 1901, edition remains uncertain, as does the question of whether the society continued after that month, possibly under another name. At any rate, during this period the local announcements columns on the third page of the *Gazette* featured over ninety items relating to the Coral Builders, and the October 29th, 1892, issue of the paper suggests that the society was likely founded that month, though under a different name, the "Acme Gas Club," due to its original aim of helping to pay the church's gas bill; that name was quickly changed to The Coral Builders Society, according to the November 26th, 1892, issue of the paper.

The Coral Builders Society is mentioned in passing in three reference and historical publications relating to the nineteenth-century history of Cleveland: Robinson, "Cleveland," 596; Davis, *Black Americans*, 122; and "Literary Societies (Black)" in *Case Western Reserve University's Encyclopedia of Cleveland History*. I am grateful to librarians and archivists at Western Reserve Historical Society, Ohio History Connection, Cleveland Public Library Special Collections, and Special Collections at Cleveland State University's Michael Schwartz Library for searching their collections for information about the Society on my behalf, though these searches have not as yet produced additional details.

8. *Cleveland Gazette*, July 8, 1893.

9. *Cleveland Gazette*, August 18, 1894. The organ must have been expensive, since seven years later the society was still donating proceeds to pay for it; or perhaps this is a new organ? *Cleveland Gazette*, March 10, 1900.

10. Broadly speaking, the goals and practices of Cleveland's Coral Builders bear out the central claim of McHenry's study of African American literary societies: these organizations signal "the efforts of free blacks in the urban North to acquire and use their literacy" against great "systematic resistance," including discriminatory laws, racial violence, and the lack of voting rights. McHenry, *Forgotten Readers*, 3. Such societies, McHenry observes, "recognized that reading was a potentially transformative activity, not only for individuals but for society as a whole," and thus they "worked to promote activism, to foster resistance, and to create citizens in black communities throughout the United States." McHenry, 3, 19. Davis usefully discusses Black literary societies in Cleveland in particular, and establishes that they served as a forum for much more than strictly literary events. Davis, *Black Americans*, 121–23. Although Spires analyzes an earlier period of Black, urban, Northern versions of economic citizenship, some of his observations seem applicable to this case, and particularly Spires's reflections on Wilson, writing as Ethiop, and James McCune Smith, writing as Communipaw. Spires, *Practice*, 121–60.

11. My profound thanks to Miami University Humanities and Social Sciences Librarian Dr. Mark Dahlquist, who first suggested that I pursue my research on the society by seeking information on particular members and was then able to locate various invaluable sources for my discussion of the society's first president, Frank Lee. Lee has not, as far as I can tell, been the subject of much scholarly work, though he does have an entry in *Oxford African American Studies Center*. Neumann, "Lee, Frank." Neuman also mentions Lee in "Black Virginians," 14–16. Lee's pension testimony is accessible online. "Testimony of Frank Lee." Lee's name appears frequently in the *Cleveland Gazette*.

12. "Testimony of Frank Lee."

13. Neumann, "Lee, Frank."

14. Spires, *Practice*.

15. Neumann, "Black Virginians," 16.

16. Qtd. in Neumann, "Black Virginians," 15.

## CHAPTER 5: RED CORAL, BLACK ATLANTIC

1. I will refer to some of these women as "mixed race Black," following Fielder's use of the term. After acknowledging that "the language available to discuss race is often problematic, and problematically limiting," Fielder explains as follows of the term *mixed-race Black people*: the modifier "mixed-race" describes "a specific and particular way of being Black" while it also "registers the fact that such people more often than not in the nineteenth-century United States (and afterward) have identified as Black rather than as white and generally did not regard the category *mixed race* as a racial designation of its own." Fielder, *Relative Races*, 10.

2. McKittrick, *Dear Science*, 135.

3. Spires, "Genealogies," 616.

4. Here I am indebted to Christina Sharpe's formulation of Black life "in the wake," which is life that is "always swept up . . . produced and determined, though not absolutely, by the afterlives of slavery," a formulation to which I will return in the Coda. Sharpe, *In the Wake*, 8. Steeve O. Buckridge, whose work I return to below, establishes the importance of red coral ornaments to Black women in Jamaica in particular. Buckridge, *Language of Dress*, 95. I consider the practice of red coral ornamentation to be what Akinwumi Ogundiran and Paula Saunders would call a "Black Atlantic ritual," one that emerged among African and African Diaspora communities in the Atlantic world; was shaped and transformed across time by their forced entanglement in "Atlantic encounters," including "commerce, commodification, slavery, Middle Passage, colonialism, and post-emancipation"; and that "does not merely recall the past but invokes, imagines, and invents it in order to create a more meaningful present and future." Ogundiran and Saunders, *Materialities*, 2, 21.

5. McKittrick, *Dear Science*, 126. Or, as McKittrick elsewhere phrases it, these "forms of diasporic livingness . . . cannot write itself-say-itself-know-itself *within* the logics of plantocratic-colonial registers." McKittrick, 145.

6. Bernstein, *Racial Innocence*, 12.

7. And indeed, as I document in chapter 1 and will further document below, white observers in different parts of the world did notice Black women wearing red coral from the early fifteenth century onward and described that practice in many forms of media including travel narratives, trade dictionaries, natural histories, and paintings.

8. Some useful reflections on how to approach ethnographic sources more ethically include Ogundiran, "Of Small Things," 431; Ogundiran and Saunders, *Materialities*, 11–14; and Brienen, "Albert Eckhout's," 236. In this chapter I piece together disparate written and visual archival sources—such as travel accounts, plantation memoirs, and paintings—that fleetingly register the importance of red coral to African-descended women within African diaspora communities from Jamaica to New Orleans. A significant challenge of producing this partial genealogy is that, to my knowledge, Black women in African diaspora communities did not create their own written or visual records of it. By drawing on the accounts of white and Black *observers* of these women, however, I follow the methods of scholars who use such ethnographic accounts as important sources of information about the lived experiences, values, and memories of African-descended peoples in the Atlantic World. Moreover, by following red coral across all of these narratives and beyond their pages into a broader Atlantic world, this chapter seeks to bring nineteenth-century US literary studies into deeper dialogue with studies of race and material culture in the African diaspora. My methods are also influenced by Joseph Roach's discussions of "circum-Atlantic memory" and "genealogies of performance" in *Cities of the Dead*, but I find Ogundiran and Saunders's "genealogies of practice" more suited to my analysis than Roach's "genealogies of performance," since I seek to name an everyday, material practice. Roach, *Cities of the Dead*, 4–7, 25–31. As Ogundiran and Saunders observe, "Practices, as actions and ideas, take place in the matrices of history, memory, antecedents, heritage, innovations, and experimentation. Therefore, to effectively understand any practice, ritual in this case, its genealogies must be reconstructed." Ogundiran and Saunders, *Materialities*, 20.

9. A notable exception is the work of Lynn Wardley, who acknowledges Stowe's colonialist perspective yet explores Stowe's fleeting awareness of African cultural practices in the novel, particularly in scenes of Dinah's kitchen. Wardley, "Relic, Fetish, Femmage." My thanks to Robert S. Levine for alerting me to this scholarship.

10. A number of scholars of the novel have identified Frowenfeld as an outsider and interpreter or stand-in for the reader, most likely a white Northern reader. See Pears, "A Speculative Reading," 117; Foote, *Regional Fictions*, 103; and Jackson, *Barriers Between Us* 76.

11. Cable, *Grandissimes*, 71, 170–71.

12. Cable, 171.

13. While Foote does not interpret this particular scene, her rich and insightful reading of the novel in relation to its historical setting on the cusp of the Louisiana Purchase deeply informs my reading, which ultimately resists some of Foote's conclusions.

14. Foote, *Regional Fictions*, 102.

15. Cable, *Grandissimes*, 120.

16. Cable, 120.

17. Cable, 121, 120.

18. Jackson, *Barriers*, 85.

19. Foote, *Regional Fictions*, 119–20.

20. This perspective on Palmyre, which Foote also suggests, would position her within the representational tradition of the "tragic mulatta," a nineteenth-century US literary trope according to which the enslaved, mixed-race Black woman is doomed to tragedy because she is neither wholly white nor Black, and is both a result of and vulnerable to sexual violence. Foote, 103. A notable example is Stowe's Eliza Harris who, like other examples of this trope in antislavery fiction, elicited readerly sympathy and raised questions about the racial identity and future of the country. For a recent study of the trope see the work of Eve Allegra Raimon who provides an overview of the literary scholarship on it. Raimon, *"Tragic Mulatta" Revisited*, 5–6.

21. Cable, *Grandissimes*, 172.

22. Cable, 120, 121.
23. In what follows I draw on variegated textual sources produced by different observers throughout the Atlantic world. These sources are not documentary, since the Black subjects depicted in these accounts had little to no control over their representation. Furthermore, heeding Ogundiran and Saunders, we must remember that there was no single "African" identity, or even single diasporic "group identity" across these disparate communities, even though their shared rituals may attest to shared "social memories of practices, places, experiences, and events." Ogundiran and Saunders, *Materialities*, 16, 20.
24. For a more detailed discussion of this practice, within these particular places within Africa, see this book's chapter 1. Here I focus more on diasporic continuances and adaptations.
25. Hartman, *Lose Your Mother*, 68. As Ogundiran observes, "the red coral bead" is one type of bead that may have preceded the Atlantic economy in Africa (via overland transport through the Sahara) and that certainly "became popular following the advent of transatlantic commerce." Ogundiran, "Of Small Things," 432, 433.
26. Buckridge, *Language of Dress*, 17.
27. Buckridge, 95.
28. Buckridge, 23. While enslaved persons in the Caribbean and North America faced significant barriers to acquiring jewelry of their preference, there were nonetheless several ways of doing so, whether by purchasing or bartering in local markets, receiving jewelry as gifts from slave traders or slave masters, possibly as incentives, or perhaps even smuggling beads aboard slave ships, though captives were typically divested of all ornaments before boarding. Buckridge, *Language of Dress*, 59; and Handler, "Middle Passage," 6–7.
29. For this interpretation of Jonkonnu I draw on Dillon, *New World Drama*, 202–7; Martinez-Ruiz, "Sketches of Memory"; Barringer, Forrester, and Martinez-Ruiz, 463–96; and Buckridge, *Language*, 98–99.
30. Dillon, *New World Drama*, 200, 202.
31. Dillon, 202.
32. Beckford, *Account*, 386. What Beckford describes is likely Jonkonnu, considering his description in light of multiple other sources, though Beckford does not name Jonkonnu.
33. Lewis, *Journal*, 74. Buckridge cites the accounts of Beckford and Lewis, though not in relation to Jonkonnu. *Language of Dress*, 58.
34. Kelly, *Voyage*, 20. Kelly claims to have spent a "residence of seventeen years in the island of Jamaica." Kelly, v.
35. See Barringer, "Catalogue: Sketches of Character," 425–61, for reproductions of Belisario's "Red Set-Girls, and Jack-in-the-Green" (item 136), "French Set-Girls" (item 141), and "Koo, Koo, or Actor-Boy" (item 139; women in background). Although no scholar, to my knowledge, has identified red coral ornaments as a key feature of Set-Girl dress, my triangulation of visual and written sources strongly suggests that it was. It is important to remember, however, that Belisario's lithographs cannot necessarily be taken as evidence that Belisario himself observed these women wearing red necklaces: as Tim Barringer and Gillian Forrester observe, Belisario's images are not "unmediated, transparent 'illustrations,' straightforwardly documentary in their implications," not least because they were mediated by many persons involved in their final production and publication. Barringer, Forrester, and Martinez-Ruiz, 1–6, 425–61.
36. Marsden, *Account*, 38–39. I believe that the women mentioned by Marsden were free, based partly on Burnard's discussions of the advantages gained by free mixed-race women who became mistresses of wealthy white Jamaican men. Burnard, *Jamaica*, 144–45.
37. Carmichael, *Domestic Manners*, 146–47. At least one source may suggest that free Black women in colonial Dutch Brazil also wore red coral: in Albert Eckhout's portrait, *African Woman and Child* (1641) (which I mention in chapter 1) the free woman of African descent wears a red coral necklace (fig. 1.5); however, as Rebecca P. Brienen reminds us, this portrait was "made for the visual consumption of someone else, generally a white male European subject." Brienen, "Albert Eckhout's," 236. And given the context and audience, it seems especially difficult to consider it as documentary evidence of anything other than the fact that Europeans associated Black women with red coral for reasons that I discuss in chapter 1.

38. Hinke, "Report," cited in Handler, "Middle Passage," 17n25. Handler notes that the "Report" does not state whether "the beads were acquired on board the slave ships or belonged to the captives before they left Africa." Handler, 17n25.

39. Gayles discusses red coral beads dating to the eighteenth century that were excavated at a plantation at White Haven, Missouri, and identified by archaeologists as objects "associated with 'Enslaved African Communities' in the area"; Gayles further presents the strong possibility that these beads were transported from Africa by enslaved persons. Gayles, "Crafting and Forging," 21–22. Anne Yentsch discusses eighteenth-century "beads of semi-precious stone or coral" excavated at a plantation in Annapolis and identifies them as "African-derived body adornment," citing traditions of red coral adornment among African and Afro-Caribbean communities. Yentsch, "Beads," 49, 50. A piece of coral, which had been buried with an African American man, was excavated from the African Burial Ground in New York City. Frohne, *African Burial Ground*, 124.

40. Potter, *Hairdresser's Experience*, 188–89.

41. Rinck's subject is likely a free woman of color, based on her tignon, a headwrap that free women of color were required to wear by city ordinance. Some scholars have speculated that the subject of this portrait is Marie Laveau (1801–81), a famed voodoo practitioner and important figure in the Creole community of New Orleans, although no known evidence suggests this, and in fact scholarly sources are firm that there is no verified image of Laveau from the period. See Long, *New Orleans Voudou Priestess*, 56–59. On the difficulties of finding reliable sources on Laveau, also see Ward. *Voodoo Queen*, xii–xiii. Some literary scholars have suggested that Laveau, in turn, was Cable's source for Palmyre Philosophe. Kennedy-Nolle, *Writing*, 208; and Sonstegard, "Graphic African," 200. However, Cable explicitly denied this, writing that he saw Laveau once "in a cottage near Congo Square, but she was in no way the model for Palmyre . . . as had been asserted in print." Turner, *Cable*, 231. For a nuanced discussion of Cable's interest in Laveau, see Long, *New Orleans Voudou*, xxvii–xxviii; and Ward, *Voodoo Queen*, 168.

42. It bears repeating that the writers under consideration in this chapter probably did not think of themselves as intentionally preserving or fostering this diasporic cultural expression.

43. Cable, *Grandissimes*, 224, 225. The narrator describes the partial "Jaloff African" ancestry of Palmyre and the "Jaloff" Bras-Coupe. Cable, 74, 224. Pears analyzes the performative nature of this scene, and offers a compelling reading of Palmyre's power and centrality in the novel more broadly. Pears, "Speculative Reading," 122.

44. Sargent, *Peculiar*, 163.

45. Sargent, 301.

46. On "pure Saxon," see Sargent, *Peculiar*, 379. For "collection of testimony," see Murison, *Politics of Anxiety*, 149.

47. For Murison's discussion of spiritual and material evidence in the novel see Murison, *Politics of Anxiety*, 139–49.

48. Sargent, *Peculiar*, 259, 197.

49. Sargent, 267–68.

50. Sargent, 273, 360–61.

51. Sargent, *Peculiar*, 368. Ultimately, as Murison observes, spiritualism "provides the missing material proof" of Clara's identity: the eponymous, formerly enslaved Peculiar Institution verifies Clara's paternity with a New Orleans medium. Yet Murison also notes that spiritualist evidence would not hold up in court. Murison, *Politics of Anxiety*, 149. But interestingly, coral does.

52. A strange comparison between coral and Clara's flesh here highlights the confusions and quandaries caused by this intense focus on color: unable to establish Clara's whiteness by reference to her skin, one of her suitors, a man of self-proclaimed abilities "to detect slight differences of color," turns to Clara's coral buttons and declares that their "delicate flesh tint" confirms them as hers. Sargent, *Peculiar*, 360. The coral's color, then, serves as proof where Clara's cannot. Sargent, 361–62. But when Clara then appears, "her lips, of a delicate coral," it seems less like proof of whiteness than of madness in endlessly weighting these "slight differences of color." Sargent, 403.

53. As Murison argues, the novel is in part Sargent's effort "to critique scientific and medical expertise." Murison, *Politics of Anxiety*, 156. Also see Alice Fahs, on how the novel shows Sargent to be "sympathetic to black freedom." Fahs, *Imagined Civil War*, 192.

54. Those "marginalized by professional medicine and science" turn to spiritualist evidence, Murison observes of the novel. Murison, *Politics of Anxiety*, 157. But they also turn to material evidence, at least in the case of coral. Murison, *Politics of Anxiety*, 157.

55. Sargent, *Peculiar*, 102, 109.

56. Sargent, 191–92, 82, 464.

57. Sargent, 197–98.

58. Sargent, 360.

59. Sargent, 368.

60. Sargent, 198, 360.

61. Sundquist, *To Wake the Nations*, 230.

62. Twain, *Pudd'nhead*, 22.

63. Twain, 32–33.

64. Twain, 45.

65. Twain, 41.

66. Twain, 160, 15, 64. My reading of Roxy as central to the novel builds on the groundbreaking essay by Carolyn Porter, "Roxana's Plot," in which Porter makes a strong case "that Roxana generates a good deal of the energy that moves the often creaky machinery of the novel forward." Porter, "Roxana's Plot," 122. Gillman focuses on the workings of another material signifier of race in the novel, the fingerprints, which Gillman reads in relation to the novel's central concern with "the problem of knowledge (social, scientific, legal) epitomized by the institution of race slavery," arguing that the book "makes us aware of . . . what George Fredrickson calls the 'official dedication to maintaining a *fictive* "race purity" for whites'." Gillman, "'Sure Identifiers,'" 89.

67. Fielder is the only scholar (to my knowledge) who finds and analyzes significant similarities between Chesnutt's short story and Twain's novel. Fielder, *Relative*, 85–118. Chesnutt's coral plot, in particular, also shares remarkable similarities with Sargent's novel: like that other Clara, this one was orphaned in childhood by a steamboat explosion on the Mississippi River near Saint Louis on the way to New Orleans with her parents. And just as in Sargent's novel, a piece of red coral jewelry worn by the infant Clara survives the accident and comes back to Clara in her adulthood, where when interpreted at exactly the right moment by a mixed-race Black woman, it verifies Clara's whiteness, as her body cannot, thereby enabling her to marry.

68. Chesnutt, "Her Virginia Mammy," 31, 34.

69. Chesnutt, 45, 49.

70. Chesnutt, 41.

71. Chesnutt, 51, 53, 55.

72. Sundquist observes that Clara is the only person who "remains under the precious social and psychological illusion" of her whiteness, though more recently Martha J. Cutter suggests, in alignment with Jennifer Riddle Harding, that "we are never quite sure what Clara understands about her racial ancestry." Cutter, "I Now Pronounce You," 198; and Harding, "Narrating the Family." For additional discussion of how Chesnutt constructs a narrative situation that erases or effaces the maternal identity of the mother, see Fielder, *Relative*, 97–102; Harding, "Narrating the Family"; and Cutter, "I Now Pronounce You," 194–99.

73. To that end Fielder's argument that race is "produced not in individual bodies themselves" but rather in gendered, nonheteronormative kinship relations between them holds particular explanatory power in any current interpretation of these works. Fielder, *Relative*, 3.

74. In the case of Chesnutt, it is more possible that he understood forms of "Black expressive culture" that Spires identifies in the work of many other nineteenth-century US Black writers. Spires, "Genealogies," 616. A number of scholars have pointed out Chesnutt's deliberate engagement with African folklore, for example, in his collection *The Conjure Woman*. Moody-Turner, "Folklore," 203, 208–10.

75. Stowe, *Cabin*, 227.

76. Stowe, 206.
77. Stowe, 207.
78. Bernstein, *Racial Innocence*, 13.
79. The material culture of the text also includes countless "Tomitudes" in the form of dolls, tobacco tins, and other items representing Stowe's characters. Merish, *Sentimental*, 152–63. See also Sarah Meer on how the text's "imagery was transferred to" multiple commodities. Meer, *Uncle Tom Mania*, 1. As a number of literary scholars have demonstrated, at this period of her life Stowe ranked people by color and other bodily features, thereby largely endorsing the classificatory schemes that determined political and legal identity and value in the antebellum United States. See Otter, "Stowe and Race"; Fredrickson, *Black Image*, 97–129; Sánchez-Eppler, *Touching Liberty*, 27–50; and Riss, *Race*, 84–110.
80. Stowe, *Cabin*, 227, 229.
81. Stowe, 229, 230.
82. Stowe, 233.
83. Stowe, 234.
84. Stowe, *A Key*, 31.
85. Stowe, *Cabin*, 234.
86. Stowe, 235.
87. Stowe, 228–29.
88. Stowe, 235.
89. Stowe, 235.
90. Stowe, 228, 229, 235.
91. Stowe, 292.
92. Stowe, 406.
93. Otter, "Stowe and Race," 24.
94. Spillers, *Black, White*, 185.
95. For example, Stephanie Foote explains that Cable's Creole characters have "no interiority" at any point in the novel. Foote, *Regional Fictions*, 120. Sean Pears says that Palmyre is granted a "sliver" of interiority in the scene with Frowenfeld that I analyze above. Pears, "Speculative Reading," 123. And Cassandra Jackson refers frequently to Palmyre as a "vessel." Jackson, *Barriers between Us*, 89.
96. Stowe, *Cabin*, 235.

## INTERCHAPTER: TOPSY IN CORAL

1. Stowe, 234.
2. Stowe, 234–35.
3. Ripa, *Iconologia*, 53.
4. Stowe, *Cabin*, 231.
5. *OED*, s.v. "Topsy (*n.*)."
6. For a discussion of this poem, see chapter 3 interchapter.

## CODA: CORAL TEMPORALITIES

1. Carson, *Edge of the Sea*, 250.
2. Carson, 207.
3. Anthropologist Stefan Helmreich, drawing on the work of Donna Haraway, reflects on the "attuning" and "weaving" powers of "the figure of the reef": this figure is now so densely populated with "durable, multiple, and porous inheritances" from multiple disciplines and cultures, he suggests, that reefs themselves "can attune their human visitors and inquisitors to empirical and

epistemological questions of scale and context," where "context" figuratively "refers to a 'weaving together,'" although precisely "which earthly and oceanic entities and agents might be woven together through reefy bones and bodies is, of course, a polysemous, shifting question." Helmreich, *Sounding the Limits*, 48, 49. For related interpretations of the epistemological, environmental, and temporal dimensions of Carson's reflections on marine materials, see Browne, *World in Which*; Alaimo, *Exposed*, 117–19, 127–28; Blum, "Bitter with the Salt"; and Navakas and Mastroianni, "Emerson Undersea."

4.  LaCapra, *Understanding Others*, 153.
5.  LaCapra, 165.
6.  In *The Mediterranean and the Mediterranean World in the Age of Philip II* (1949), Fernand Braudel famously articulates what he later calls "the plurality of temporalities" and the need for historians to attend to these. Braudel, 243. On the adaptability of the concept of a layered temporality, particularly for scholars of US literature and culture, see Susan Gillman's reflections on Braudel and, more recently, on the work of Amy Kaplan. Gillman, "Oceans" and "The Political," 176. Relatedly, drawing on Braudel, Glissant, Gilroy, and Benjamin in his analysis of the intertwined and ongoing logics and violence of capitalism and slavery, Ian Baucom usefully observes that "time does not pass, it accumulates." Baucom, *Specters of the Atlantic*, 24. Below I discuss other scholars who engage with a layered temporality for different purposes.
7.  LaCapra, 156–57.
8.  Trouillot, *Silencing the Past*, 49; Hartman, *Lose Your Mother*, 6; and Jenkins and Leroy, *Histories of Racial Capitalism*, 12.
9.  LaCapra, *Understanding Others*, 165; and Fleissner, "Historicism Blues," 711.
10. Hale, "Coral Branch," 16–20.
11. Latimer, "Coral Creations," 117.
12. DeMond, "Negro Element," 22.
13. DeMond, 12, 9.
14. DeMond, 9, 10.
15. DeMond, 12, 10.
16. DeMond, 12.
17. Here DeMond prefigures some key aspects of the more recent project of racial capitalism.
18. As Juliana Chow usefully observes, reflecting on the difference between popular understandings of coral today and of the period before 1900, "coral, today, is an ominous figure, associated with loss and extinction." Chow, *Nineteenth-Century American Literature*, 101. An excellent tool for understanding, visualizing, predicting, preparing for, and responding to the stressors endangering coral reef ecosystems is the National Oceanic and Atmospheric Administration's "Coral Reef Watch," which has been gathering and publicly disseminating data for over twenty years. https://coralreefwatch.noaa.gov/.
19. Melville, *Moby-Dick*, 17.
20. Michelle C. Neely insightfully explores the environmental implications of this metaphor, as it is developed and contested in the work of Melville and of A. S. Byatt. Neely, *Against Sustainability*, 103–13. Stacy Alaimo explores the damaging ecological effects of metaphors of the sea as a space that is "alien" to human affairs and analyzes Carson's writings for a potential challenge to that perspective, though a challenge that is not without its own flaws. Alaimo, *Exposed*, 113, 117–19, 127–28.
21. Jones delivered these remarks in an as-yet-unpublished conference paper, "Weathering, Wreck, Salvage: C19 Oceanic Time and the Metaphorical Turn in Oceanic Studies," given at the *Society of Nineteenth-Century Americanists* conference in March 2022.
22. Sharpe, *In the Wake*, 21. In connection with this idea, see also Katherine McKittrick on the "black Atlantic livingness" of M. NourbeSe Philips's *Zong!*. *Dear Science*, 126; 125–150; and Joshua Bennett's analysis of Black writing on oceans and ocean life. Bennett, *Being Property Once*. 169–89.
23. For this framing, and for giving me the chance to think through these ideas at an earlier stage, I thank Lauren LaFauci, who organized the panel, "Before the Anthropocene? Placing Early Amer-

ica in Environmental Humanities," a Society of Early Americanists-sponsored Panel at ASLE (Association for the Study of Literature and Environment), held at the University of California-Davis in 2019.

24. Neely, *Against Sustainability*, 17.
25. Fleissner, "Historicism Blues," 702.
26. Tamarkin, "Reading for Relevance."

# BIBLIOGRAPHY

PRIMARY SOURCES (WITH ORIGINAL PUBLICATION DATES)

Agassiz, Elizabeth Cabot Cary. *A First Lesson in Natural History*. Boston: Little, Brown, 1859.
———."The World We Live On: What Are Corals?" *Our Young Folks: An Illustrated Magazine for Boys and Girls*, April 1869, 217–26.
Akin, James, Samuel Allardice, William Barker, Joseph Bowes, Thomas Clarke, William Creed, Benjamin Jones, Alexander Lawson, William Ralph, Robert Scot, Francis Shallus, James Smither, James Thackara, James Trenchard, John Vallance, and Henry Weston. "Coral Fishery." In *Encyclopaedia; Or, a Dictionary of Arts, Sciences, and Miscellaneous Literature*, 5:443. Philadelphia: Printed by Thomas Dobson, at the stone house, no. 41, South Second Street, 1798.
Alcott, Louisa May. "The Whale's Story." 1867. In *Morning-glories, and Other Stories*, 135–45. New York: Carleton, 1871.
Andrews, Jane. "Sea-Life." *Our Young Folks: An Illustrated Magazine for Boys and Girls*, January 1866, 331–38.
———. *The Stories Mother Nature Told Her Children* 1888. Boston: Ginn & Company, Publishers, 1893.
Anna. "The Coral Insect." *The Operatives' Magazine*, March 1842, 191.
Aveling, Edward Bibbens. *The Student's Darwin*. London: Free Thought, 1881.
Barnett, Ferdinand Lee. "Race Unity—Its Importance and Necessity." 1879. In *Lift Every Voice: African American Oratory, 1787–1900*, edited by Philip S. Foner and Robert James Branham, 603–7. Tuscaloosa: University of Alabama Press, 1998.
Barrera, Madame de. *Gems and Jewels, Their History, Geography, Chemistry, and Analysis, from the Earliest Ages Down to the Present Time*. London: Richard Bentley, 1860.
Barrow, John. "Coral Fishery." In *Dictionarium polygraphicum: or, The Whole Body of Arts Regularly Digested [. . .] Adorned with Proper Sculptures, Curiously Engraven on More than Fifty Copper Plates*, 1:197–98. London: Printed for C. Hitch and C. Davis in Pater-Noster Row, and S. Austen in St. Paul's Church-Yard, 1735.
Beckford, William. *A Descriptive Account of the Island of Jamaica: With Remarks upon the Cultivation of the Sugar-Cane, throughout the Different Seasons of the Year, and Chiefly Considered in a Picturesque Point of View: Also Observations and Reflections upon What Would Probably Be the Consequences of an Abolition of the Slave-Trade, and of the Emancipation of the Slaves*. Vol. 2. London: Printed for T. and J. Egerton, 1790.
Beecher, Henry Ward. "Christianity a Vital Force." In *Sermons by Henry Ward Beecher, Plymouth Church, Brooklyn*, 2:301–20. New York: Harper, 1868.
———. "Sermon." *The Independent [. . .] Devoted to the Consideration of Politics, Social and Economic Tendencies, History, Literature, and the Arts*, August 20, 1863, 2.
———. "The Unity of Man." In *The Sermons of Henry Ward Beecher in Plymouth Church, Brooklyn. From Verbatim Reports by T.J. Ellinwood*. "Plymouth Pulpit," Seventh Series: September, 1871–March, 1872, 195–216. New York: J. B. Ford & Company, 1872.
Best, Mrs. M. C. *The Works of Creation Illustrated*. Philadelphia: Presbyterian Board of Publication, 1840.
Blake, John Lauris. *The Parlor Book; or, Family Encyclopedia of Useful Knowledge and General Literature*. New York: J. L. Piper, 1837.
———. *The Wonders of the Ocean: Containing an Account of the Color, Saltness, and Probable Depth of the Ocean; of Its Mountains of Ice, Gulfs, Whirlpools, Currents, and of Its Coral and Other Formations and Productions, so Far as Known, whether Animal, Mineral, or Vegetable*. Cazenovia, NY: Henry and Sweetlands, 1845.

Bosman, Willem. *A New and Accurate Description of Guinea*. London: J. Knapton, 1705.

Boyle, Robert. *The Origine of Forms and Qualities*. 1666. In *Selected Philosophical Papers of Robert Boyle*, edited with an introduction by M. A. Stewart, 1–96. Indianapolis: Hackett Publishing Company, 1991.

"A Branch of Coral." *Friends Intelligencer* 17, no. 26 (1860): 415.

Brown, Benjamin Gratz. *Speech of the Hon. B. Gratz Brown, of St. Louis, on the Subject of Gradual Emancipation in Missouri: Delivered in the House of Representatives, February 12, 1857*. Saint Louis: Printed at Missouri Democrat Book and Job Office, 1857.

Bryant, William Cullen, ed. *A Library of Poetry and Song, Being Choice Selections from the Best Poets*. New York: J. B. Ford and Company, 1871.

Buffon, Georges Louis Leclerc. *Barr's Buffon. Buffon's Natural History. Containing A Theory of the Earth, A General History of Man, of the Brute Creation, and of Vegetables, Minerals, &c. &c.*. Vol. 1. London, 1797.

"The Busy Coral Insect." *New York Times*, December 8, 1889.

Cable, George Washington. *The Grandissimes*. New York: Scribner's, 1880.

Carmichael, Mrs. A. C. *Domestic Manners and Social Condition of the White, Coloured and Negro Population of the West Indies. By Mrs. Carmichael [. . .] In two volumes*. London: Whittaker, Treacher, 1833.

Carson, Rachel. *The Edge of the Sea*. 1955. Boston: Mariner Books, 1998.

Catesby, Mark. *Natural History of Carolina, Florida, and the Bahama Islands*. Vol. 1. London: Printed at the expence of the author, and sold by W. Innys and R. Manby, at the West End of St. Paul's, by Mr. Hauksbee, at the Royal Society House, and by the author, at Mr. Bacon's in Hoxton, 1731.

Chambon, M. *Le Guide du commerce de l'Amérique*. Marseille: A. Avignon, 1777.

Chesnutt, Charles W. "Her Virginia Mammy." In *The Wife of His Youth*, 25–59. Boston: Houghton, Mifflin, 1899.

"Children's 'Morning Star' Mission." *Missionary Herald*, August 1886, 320.

Church, William C. "A Midwinter Resort: With Engravings of Winslow Homer's Water-Color Studies in Nassau." *Century Illustrated Magazine* (1887): 499–507.

Conneau, Théodore, and Brantz Mayer, eds. *Adventures of an African Slaver: An Account of the Life of Captain Theodore Canot, Trader in Gold, Ivory, and Slaves on the Coast of Guinea*. New York: D. Appleton, 1854.

"A Conversation about Islands." *Merry's Museum and Parley's Magazine*, July 1, 1853, 70–75.

"The Coral." *Harbinger, Devoted to Social and Political Progress*, December 1845, 35.

"Coral." *Robert Merry's Museum* 14, no. 1 (July 1847): 2–3.

"Coral and Coral Islands." *The Friend: A Religious and Literary Journal* 8, no. 36 (June 1835): 281–82.

"The Coral Insect." *Christian Register* 25, no. 39 (September 1846): 156.

"The Coral Insect: Original." *The Boston Pearl, a Gazette of Polite Literature. Devoted to Original Tales, Legends, Essays, Translations, Travelling, Literary and Historical Sketches, Biography, Poetry, Criticisms, Music, etc.* 5, no. 13 (December 1835): 99.

"Coral Islands." *The Museum of Foreign Literature, Science, and Art* 2, no. 11 (May 1823): 394–400.

"Coral Reefs." *Godey's Lady's Book* 56, no. 3 (1858): 239–40.

"The Coral Workers." *Advocate and Family Guardian* 38, no. 7 (April 1872): 78–79.

"Cousin Annie's Riddle." *Youth's Penny Gazette*, February 2, 1848, 12.

Coxe, Margaret. *Claims of the Country on American Females*. 2 vols. Columbus, OH: I. N. Whiting, 1842.

———. *The Wonders of the Deep; or, Two Months at the Seashore*. New York: Published by the General Protestant Episcopal Sunday School Union, 1836.

Dana, James Dwight. *Corals and Coral Islands*. New York: Dodd and Mead, 1872.

———. "From Professor Dana: Montgomery's 'Pelican Island,' New Haven, May 27, 1853." In *The Life of James Dwight Dana: Scientific Explorer, Mineralogist, Geologist, Professor in Yale University*, by Daniel Coit Gilman, 212–13. New York: Harper and Brothers, 1899.

———. *On Coral Reefs and Islands*. New York: Putnam, 1853.

———. *Synopsis of the Report on Zoophytes of the US Exploring Expedition*. New Haven, CT: Published by the author, 1859.

———. *United States Exploring Expedition*. Vol. 7, *Zoophytes*. Philadelphia: C. Sherman, 1846.

———. *United States Exploring Expedition*. Vol. 10, *Geology*. Philadelphia: C. Sherman, 1849.

———. *Zoophytes: Atlas*. Philadelphia: Lea and Blanchard, 1849.

Dapper, Olfert. *Naukeurige Beschrijvingen der Afrikaensche Gewesten*. Amsterdam: Jacob van Meurs, 1668.

Darwin, Charles. *On the Structure and Distribution of Coral Reefs; Also Geological Observations on the Volcanic Islands and Parts of South America Visited during the Voyage of H.M.S. Beagle. with Maps Plates and Numerous Illustrations and a Critical Introduction to Each Work by Prof. John W. Judd*. Edited by John Wesley Judd. London: Ward, Lock, 1890.

———. *The Structure and Distribution of Coral Reefs*. London: Smith, Elder, 1842.

Darwin, Erasmus. *The Botanic Garden: A Poem, in Two Parts*. Part 1, *Containing the economy of vegetation*. Part 2, *The loves of the plants*. With philosophical notes. London: Printed for J. Johnson, St. Paul's Church Yard, 1791.

Davis, Rebecca Harding. "Life in the Iron-Mills." 1861. In *Life in the Iron-Mills*, edited by Cecelia Tichi, 39–74. Boston: Bedford Cultural Editions, 1998.

"A Day on a Coral Reef." *Christian Advocate and Journal*, June 23, 1864, 194.

"A Day on a Coral Reef." *The Friend: a Religious and Literary Journal*, July 30, 1864, 377.

DeMond, Abraham Lincoln. *The Negro Element in American Life. An Oration Delivered by Rev. A. L. DeMond, in the Dexter Avenue Baptist Church, Montgomery, Alabama, Jan. 1, 1900*. Montgomery: Alabama Printing Company, 1900.

Des Bruslons, Jacques Savary. *Dictionnaire universel de commerce*. Vol. 1. Paris: Veuve Estienne, 1741.

Diderot, Denis. "Coral." In *The Encyclopedia of Diderot & d'Alembert Collaborative Translation Project*, translated by Eugene Navakas, 4:197. Ann Arbor: Michigan Publishing, University of Michigan Library, 2018. http://hdl.handle.net/2027/spo.did2222.0003.460.

Diderot, Denis, and Gabriel-François Venel. "Coral." In *The Encyclopedia of Diderot and d'Alembert Collaborative Translation Project*, translated by Eugene Navakas, 4:196–97. Ann Arbor: Michigan Publishing, University of Michigan Library, 2018. http://hdl.handle.net/2027/spo.did2222.0003.462.

Diderot, Denis, and Louis-Jean-Marie Daubenton. "Coral." In *The Encyclopedia of Diderot & d'Alembert Collaborative Translation Project* translated by Eugene Navakas, 4:194–96. Ann Arbor: Michigan Publishing, University of Michigan Library, 2018. http://hdl.handle.net/2027/spo.did2222.0002.366.

Ellis, John. *An Essay Towards a Natural History of the Corallines: And Other Marine Productions of the Like Kind, Commonly Found on the Coasts of Great Britain and Ireland; To Which is Added the Description of a Large Marine Polype Taken Near the North Pole, by the Whale-Fishers, in the Summer 1753*. London: Printed for the Author, 1755.

Emerson, Ralph Waldo. "Experience." 1844. In *Essays and Lectures*, edited by Joel Porte, 469–92. New York: Library of America, 1983.

———. Ralph Waldo Emerson to Lidian Emerson, January 14, 1843, Barnum's Hotel, Baltimore, MD. In *The Letters of Ralph Waldo Emerson 1842–1847*, edited by Ralph L. Rusk, 123–24. Vol. 3 of *The Letters of Ralph Waldo Emerson*. New York: Columbia University Press, 1941.

Flinders, Matthew. *A Voyage to Terra Australis: Undertaken for the Purpose of Completing the Discovery of That Vast Country, and Prosecuted in the Years 1801, 1802, and 1803*. Vol. 2. London: Bulmer, 1814.

"For the Coral Workers." *Life and Light for Woman*, March 1888, 154–55.

"For Next Christmas." *Woman's Work for Woman*, October 1871, 148.

"Formation of New Islands." *New Harmony Gazette*, September 3, 1828, 359.

Fuller, Margaret. Margaret Fuller to William H. Channing, October 25 and 28, 1840. In *1839*, ed. Robert N. Hudspeth, 170–78. Vol. 2 of *The Letters of Margaret Fuller*. Ithaca, NY: Cornell University Press, 1983.

Gilman, B. F. *Drops of Water; Or, Pencillings by the Seashore*. Boston: Massachusetts Sabbath School Society, 1856.

Goodrich, Samuel Griswold. *Peter Parley's Wonders of the Earth, Sea, and Sky*. New York: S. Colman, 1840.

Grandpré, L. de (Louis). *Voyage à la côte occidental d' Afrique*. Paris: Dentu, imprimeur-libraire, Palais du Tribunat, galeries des bois, 1801.

"The Greatness of Little Things." *Friends' Intelligencer*, May 26, 1855, 159–60.

H. T. C. "A Paper on Corals." *Christian Parlor Magazine*, March 1, 1848, 333–34.

Hale, Sarah Josepha Buell. "The Coral Branch." In *The School Song Book: Adapted to the Scenes of the School Room, Written for American Children and Youth*, 16. Boston: Allen and Ticknor, 1834.

———. "Coral Reefs." *Godey's Lady's Book* 56 (1858): 239–40.

———. "The General Resemblance of Vegetable and Animal Life." In *Good's Book of Nature. Abridged from the Original Work. Adapted to the Reading of Children and Youth, with Questions for the Use of Schools and Illustrations from Original Designs*, 40–45. Boston: William D. Ticknor, 1843.

Hamlin, Augustus Choate. *Leisure Hours Among the Gems*. Boston: J. R. Osgood, 1884.

Harper, Frances Ellen Watkins. *Poems*. Philadelphia: Merrihew and Son, 1871.

Haughwout, Frank G. "The Coral Industry in Italy. Report by Consul Haughwout, of Naples." In *Reports from the Consuls of the United States on the Commerce, Manufactures, Etc., of their Consular Districts*, 120–21. Washington, DC: Government Printing Office, 1883.

Hicks, Josephus Franklin, Ellen E. Hicks, and Enid E. Callahan. *St. John A.M.E. Church Yesterday and Today: A Descriptive Historical Record of the Development of St. John A.M.E. Church in Cleveland, Ohio*. Cleveland, 1956.

Hickson, Sydney J. *An Introduction to the Study of Recent Corals*. Manchester: Manchester University Press, 1924.

Hinke, W. J., ed. "Report of the Journey of Francis Louis Michel from Berne, Switzerland, to Virginia, October 2, 1701–December 1, 1702." *Virginia Magazine of History and Biography* 24 (1916): 1–43.

"Historical Articles Related to the St. John African Methodist Episcopal Church." Unpublished document created by St. John A.M.E. Paper copy.

"The History of St. John African Methodist Episcopal Church, Cleveland, Ohio." Unpublished document created by St. John A.M.E.

Holder, C. F. "Un-Natural History." *Outlook*, November 23, 1895, 844–45.

Honeyman, Mary F. "Sea-Gardens off the Bermudas." *The Chautauquan; A Weekly Newsmagazine*, June 1895, 340–42.

Hughes, Griffith. *The Natural History of Barbados*. London: Printed for the Author, 1750.

Humboldt, Alexander von. *Views of Nature*. 1808. Edited and translated by Stephen T. Jackson, Laura Dassow Walls, and Mark W. Person. Chicago: University of Chicago Press, 2016.

Huxley, Thomas Henry. "On Coral and Coral Reefs—A Lecture, Delivered in the Hulme Town Hall, Manchester, November 4th, 1870." In *Critiques and Addresses*, 113–33. New York: D. Appleton and Company, 1882.

Jacobs, Harriet. *Incidents in the Life of a Slave Girl*. 1861. Edited by Tiya Miles. New York: Modern Library, 2021.

"James Montgomery." *Eclectic Magazine of Foreign Literature*, August 1854, 455–64.

Jamieson, Robert. *Commerce with Africa*. London: Effingham Wilson, 1859.

Johnston, James F. W. "The Circulation of Matter." *Blackwood's Magazine*, May 1853, 550–60.

Kelly, James. *Voyage to Jamaica, and Seventeen Years' Residence in that Island*. Belfast: Printed by J. Wilson, 1838.

Kidder, Daniel Parish. *The Coral-Maker*. New York: Sunday School Union of the Methodist Episcopal Church, 1844.

Knight, Charles. "On Coral and the Coral Fishery." 1840. *The Family Magazine; Or Monthly Abstract of General Knowledge*, 1841, 342–43.

"Labors of the Coral." *Pittsburgh Legal Journal* 1, no. 32 (1853): 1.

Lander, John, and Richard Lander. *Journal of an Expedition to Explore the Course and Termination of the Niger*. London: John Murray, 1832.

Landolphe, J. F. *Mémoires du Capitaine Landolphe, contenant l'histoire de ses voyages pendant trente-six ans, aux côtes d'Afrique et aux deux Amériques*. Vol. 1. Paris: Arthus Bertrand, 1823.

Latimer, Elizabeth Wormeley. "Coral Creations." *Illustrated Waverley Magazine and Literary Repository: Devoted to Original Tales, Poetry, Music and Choice Miscellaneous Reading*, February 23, 1856, 117.

Lewis, Matthew Gregory. *Journal of a West-India Proprietor, Kept During a Residence in the Island of Jamaica*. London: John Murray, 1834.

"Literary: The Pelican Island." *Ladies Garland*, October 6, 1827, 67.

"Literature, &C.: Montgomery's New Poem." *The Albion: A Journal of News, Politics and Literature* 6, no. 14 (September 1827): 105–6.

"Little Builders." *Juvenile Missionary Magazine*, June 1846, 143–44.

"Little Builders." *Zion's Herald and Wesleyan Journal*, June 30, 1847, 104.

Longfellow, Henry Wadsworth. "To a Child." In *The Belfry of Bruges and Other Poems*, 51–61. Cambridge: John Owen, 1845.

Mann, Horace. "An Historical View of Education, Showing its Dignity and its Degradation." 1845. In *Lectures, and Annual Reports, on Education*, 215–68. Cambridge: Published for the Editor, 1867.

Mann, Mary Peabody. "Coral." *Arthur's Home Magazine*, January 1880, 575–77.

Marsden, Peter. *An Account of the Island of Jamaica*. Newcastle: Printed for the Author by S. Hodgson, 1788.

Marsilli, Louis Ferdinand comte de. *Histoire physique de la mer*. 1725. Translated by Anita McConnell. Bologna: Museo di fisica dell'Università di Bologna, 1999.

Marx, Karl. *Capital, Volume 1*. 1867. Vol. 35, *The Collected Works of Marx and Engels, Economic Works 1857–1894*. New York: International Publishers, 1975.

Maury, Matthew Fontaine. *The Physical Geography of the Sea*. London: Sampson Low, Son, & Co., 1855.

Melville, Herman. *Moby-Dick; Or, The Whale*. 1851. Edited by Hester Blum. New York: Oxford University Press, 2022.

Miller, Samuel. *A Brief Retrospect of the Eighteenth Century: Part First; In Two Volumes: Containing a Sketch of the Revolutions and Improvements in Science, Arts, and Literature, During That Period; By Samuel Miller, A. M. One of the Ministers of the United Presbyterian Churches in the City of New-York, Member of the American Philosophical Society, and Corresponding Member of the Historical Society of Massachusetts*. 2 vols. New York: Printed by T. and J. Swords, 1803.

Mitchell, Ellen M. "Corals." *Arthur's Illustrated Home Magazine*, December 1873, 756–58.

Montgomery, James. *The Pelican Island, and Other Poems*. Philadelphia: E. Littel and J. Grigg, 1827.

"Montgomery's Pelican Island—Prof. Dana's Criticisms." *Norton's Literary Gazette*, June 15, 1853, 98–99.

Moody, Sophy. "The Story of a Coral Bracelet." In *The Fairy Tree; or, Stories from Far and Near*, 30–41. London: T. Nelson and Sons, 1861.

Moore, Rachel Wilson. *Journal of Rachel Wilson Moore Kept During a Tour to the West Indies and South America, in 1863–64*. Philadelphia: T. E. Zell, 1867.

Morton, Samuel George. *Crania Americana*. Philadelphia: J. Dobson, 1839.

Nichols, Grace. *I Is a Long Memoried Woman*. London: Karnak House, 1983.

Nimmo, William P. *Omens and Superstitions: Curious Facts and Illustrative Sketches*. Edinburgh: W. P. Nimmo, 1868.

Nott, Josiah Clark, George Robbins Gliddon, and Samuel George Morton. *Types of Mankind, Or, Ethnological Researches: Based Upon the Ancient Monuments, Paintings, Sculptures, and Crania of Races, and Upon Their Natural, Geographical, Philological, and Biblical History*. Philadelphia: J. B. Lippincott, 1854.

Ogilby, John. *Africa [. . .]*. London: John Ogilby, 1670.

"Original Pencil Sketches: Salmagundi; Or Paragraphs from the Common Place Book of John Greenleaf Buckeye. VII. The Coral Insect." *Cincinnati Mirror, and Western Gazette of Literature, Science, and the Arts* 4, no. 38 (July 1835): 305.

Orwell, George. *Nineteen Eighty-Four*. 1949. New York: Penguin, 1981.

Ovid. *Metamorphoses*. Translated by A. S. Kline. In The Ovid Collection, University of Virginia Electronic Text Center. https://ovid.lib.virginia.edu.

Peale's Museum (Philadelphia, Pa.). *Guide to the Philadelphia Museum*. Philadelphia: Museum Press, 1805.

Peyssonnel, Jean-André, and William Watson. "An Account of a Manuscript Treatise, Presented to the Royal Society, Entitled *Traité du Corail*." *Philosophical Transactions* 47 (1751–52): 445–69.

Phaer, Thomas, and Rick Bowers, eds. *Thomas Phaer and the Boke of Chyldren*. 1544. Tempe: Arizona Center for Medieval and Renaissance Studies, 1999.

Phillips, Thomas. *A Journal of a Voyage Made in the Hannibal of London, Ann. 1693, 1694 from England to Cape Monseradoe, in Africa*. In *A Collection of Voyages and Travels [. . .] in Six Volumes*, compiled by Awnsham Churchill and John Churchill, 6: 172–239. London: Printed by assignment from Messrs. Churchill for John Walthoe, 1732.

Pliny. *The Natural History of Pliny*. Translated by John Bostock and Henry T. Riley. London: Henry G. Bohn, 1855.

Poe, Edgar Allan. "Instinct vs Reason—A Black Cat." 1840. In *Poetry and Tales*, edited by Patrick F. Quinn, 370–72. New York: Library of America, 1984.

Potter, Eliza. *A Hairdresser's Experience in the High Life*. 1859. New York: Oxford University Press, 1991.

Prince, Mary. *The History of Mary Prince, a West Indian Slave*. 1831. In *Six Women's Slave Narratives*, edited by William Andrews, 1–40. New York: Oxford University Press, 1988.

Quin, John T. *The Building of an Island: Being a Sketch of the Geological Structure of the Danish West Indian Island of St. Croix, or Santa Cruz*. New York: Chauncey Holt, 1907.

Rankine, Claudia. *Citizen: An American Lyric*. Minneapolis, MN: Graywolf Press, 2014.

———. "'The Condition of Black Life Is One of Mourning': On Racial Violence." *New York Times*, June 22, 2015.

"Review of Books." *Christian Lady's Magazine*, November 1842, 467.

Ripa, Cesare. *Iconologia; Or, Moral Emblems*. London: Printed by Benjamin Motte, 1709.

Sargent, Epes. *Peculiar: A Tale of the Great Transition*. New York: Carleton, 1864.

Schele de Vere, Maximilian. *Wonders of the Deep: A Companion to Stray Leaves from the Book of Nature*. New York: G. P. Putnam and Son, 1869.

"The Science of Geology [from the Glasgow Treatises, with Additions] [. . .]." *Family Magazine; Or, Monthly Abstract of General Knowledge*, May 1, 1838, 405–9.

"The 'Science' of Poetry—Montgomery's Pelican Island." *Norton's Literary Gazette*, May 15, 1853, 82.

Sigourney, Lydia Howard Huntley. "The Coral Insect." In *Poems*, 16–17. Boston: S. G. Goodrich, 1827.

Simmonds, Peter Lund. *The Commercial Products of the Sea: Marine Contributions to Food, Industry, and Art*. London: Griffith and Farran, 1879.

"A Sixth Continent." *Western Luminary*, July 5, 1826.

Slade, Mary B. C. "The Coral Insect, or Perseverance." In *Merry Chimes: A Collection of Songs, Duets, Trios, and Sacred Pieces, for Juvenile Classes, Public Schools, and Seminaries. To Which Is Prefixed Complete Elementary Instructions, and Attractive Exercises*, by L. O. Emerson, 46. Boston: Oliver Ditson, 1865.

"Small Efforts Well Directed." *Orphans' Advocate and Social Monitor* 9, no. 5 (1850): 36.

Smith, James McCune. "Civilization: Its Dependence on Physical Circumstances." *The Anglo-African Magazine*, January 1859. Reprinted in *The Works of James McCune Smith: Black Intellectual and Abolitionist*, edited by John Stauffer, 246–63. New York: Oxford University Press, 2006.

———. James McCune Smith to Gerrit Smith, May 12, 1848. In *The Works of James McCune Smith: Black Intellectual and Abolitionist*, edited by John Stauffer, 307–9. New York: Oxford University Press, 2006.

Smith, Roswell C. *Smith's New Geography: Containing Map Questions Interspersed with Such Facts As an Observing Tourist Would Notice, Which Are Followed by a Concise Text and Explanatory Notes; Based on a Combination of the Analytical, Synthetical and Comparative Systems, Designed to Be Simple and Concise, but Not Dry, Philosophical, yet Practical; For the Use of Common Schools in the United States and Canada*. Philadelphia: J. B. Lippincott, 1860.

Spallanzani, Lazzaro. *Travels in the Two Sicilies: And Some Parts of the Apennines. Translated from the original Italian of the Abbe Lazzaro Spallanzani*. London: Printed for G. G. and J. Robinson, 1798.

Spofford, Harriet Prescott. "A Mutual Dependence." In *Stepping-Stones to Happiness*, 65–68. New York: Bible House, 1897.

Stark, James Henry. *Stark's History and Guide to the Bahama Islands*. Boston: James H. Stark, 1891.

Stowe, Harriet Beecher. "The Coral Ring." In *The Christian Souvenir: An Offering for Christmas and The New Year*, edited by Isaac F. Shepard, 265–81. Boston: H. B. Williams, 1843.

———. *A Key to "Uncle Tom's Cabin."* Boston: John P. Jewett and Company, 1853.

———. *Uncle Tom's Cabin*. 1852. Edited by Elizabeth Ammons. New York: W. W. Norton, 2018.

Taylor, Helen. *Missionary Hymns for the Use of Children*. London: Hamilton, Adams, 1846.

Tenney, Abby Amy (Gove). *Pictures and Stories of Animals: For the Little Ones at Home*. New York: Sheldon and Company, 1868.

"Testimony of Frank Lee: November 12, 1903." Black Virginians in Blue, Nau Center for Civil War History, the University of Virginia. Accessed April 20, 2022. http://community.village.virginia.edu /usct/node/521.

Thompson, Joseph P. *The Moral Unity of the Human Race*. New York: M. W. Dodd, 1851.

Thompson, Mrs. "The Coral Worm." *Rochester Gem and Ladies' Amulet* 9, no. 1 (1836): 5.

Tocque, Philip. *The Mighty Deep*. New York: Carleton and Porter, 1852.

Townsend, Jonas Holland. "The Policy That We Should Pursue." *Anglo-African Magazine*, October 1859, 324–27.

Twain, Mark. *The Tragedy of Pudd'nhead Wilson: And the Comedy, Those Extraordinary Twins*. 1894. New York: Oxford University Press, 1996.

U. F. "A Talk About Coral." In *Woodworth's Youth's Cabinet*, 14–17. New York: D. Austin Woodworth, 1856.

"Virginia." *Christian Observer* 94, no. 50 (December 1906): 13.

Voltaire. "Du Corail." In *Les Singularités de la nature*, 8. Basel, 1768.

Wakefield, Priscilla. *Mental Improvement: Or, The Beauties and Wonders of Nature and Art; In a Series of Instructive Conversations; By Priscilla Wakefield, Author of "Leisure Hours."* New Bedford, MA: Printed by Abraham Shearman, Jun. for Caleb Greene and Son, 1799.

Warfield, Catherine Anne. *Miriam Montfort: A Novel*. New York: D. Appleton and Company, 1873.

Watson, Henry Cood. "The Coral and Bell." *Every Youth's Gazette* 1, no. 3 (1842): 24.

Webb, George James. "The Coral Branch." *American Annals of Education and Instruction: Being a Continuation of the American Journal of Education* 5, no. 3 (1835): 144.

"Wonders of the Deep: Coral and the Coral Makers." *Eclectic Magazine of Foreign Literature*, March 1856, 371.

"A Word of Encouragement." *Spirit of Missions* 62, no. 10 (October 1897): 570–72.

"Works of the Coral Insect." *Arthur's Home Magazine*, May 1, 1868, 302.

Wright, Anne. *The Observing Eye: Or, Letters to Children on the Three Lowest Divisions of Animal Life: The Radiated, Articulated, and Molluscous*. London: Jarrold and Sons, 1851.

## SECONDARY SOURCES

Abiodun, Rowland. *Yoruba Art and Language: Seeking the African in African Art*. Cambridge: Cambridge University Press, 2014.

Adkison, Jennifer Dawes. "Elizabeth Agassiz." In *Early American Nature Writers: A Biographical Encyclopedia*, edited by Daniel Patterson, Roger Thompson, and J. Scott Bryson, 8–13. Westport, CT: Greenwood Press, 2008.

Alaimo, Stacy. *Exposed: Environmental Politics and Pleasures in Posthuman Times*. Minneapolis: University of Minnesota Press, 2016.

———. *Undomesticated Ground: Recasting Nature as Feminist Space*. Ithaca, NY: Cornell University Press, 2000.

Anderson, Alicia K., and Lynn A. Price, eds. *George Washington's Barbados Diary, 1751–2*. Charlottesville: University of Virginia Press, 2018.

Anderson, Katharine. "Coral Jewellery." *Victorian Review* 34, no. 1 (2008): 47–52.

Ball, Berenice. "Whistles with Coral and Bells." *Magazine Antiques* 80 (December 1961): 552–55.

Baptist, Edward. *The Half Has Never Been Told: Slavery and the Making of American Capitalism*. New York: Basic Books, 2014.

Barrett, Lindon. *Conditions of the Present: Selected Essays*. Edited by Janet Neary with contributions by Elizabeth Alexander, Jennifer DeVere Brody, Daphne A. Brooks, Linh U. Hua, Marlon B. Ross, and Robyn Wiegman. Durham, NC: Duke University Press, 2018.

———. "Identities and Identity Studies: Reading Toni Cade Bambara's 'The Hammer Man.'" In *Conditions of the Present: Selected Essays*, edited by Janet Neary, 171–92. Durham, NC: Duke University Press, 2018.

Barringer, Tim, Gillian Forrester, and Barbaro Martinez-Ruiz, eds. *Art and Emancipation in Jamaica: Isaac Mendes Belisario and His Worlds*. New Haven, CT: Yale University Press, 2007.

Baucom, Ian. *Specters of the Atlantic: Finance Capital, Slavery, and the Philosophy of History*. Durham, NC: Duke University Press, 2005.

Bauer, Ralph, and José Antonio Mazzotti, eds. *Creole Subjects in the Colonial Americas: Empires, Texts, Identities*. Chapel Hill: University of North Carolina Press, 2009.

Baym, Nina. *American Women of Letters and the Nineteenth-Century Sciences: Styles of Affiliation*. New Brunswick, NJ: Rutgers University Press, 2002.

———. "Reinventing Lydia Sigourney." *American Literature* 62, no. 3 (September 1990): 385–404.

Beckert, Sven, and Seth Rockman, eds. *Slavery's Capitalism: A New History of American Economic Development*. Philadelphia: University of Pennsylvania Press, 2016.

Beer, Gillian. *Open Fields: Science in Cultural Encounter*. Oxford: Clarendon Press, 1996.

Bellion, Wendy. *Citizen Spectator: Art, Illusion, and Visual Perception in Early National America*. Chapel Hill: University of North Carolina Press, 2011.

Ben-Amos, Paula. *The Art of Benin*. Rev. ed. Washington, DC: Smithsonian Institution Press, 1995.

Bennett, Joshua. *Being Property Once Myself: Blackness and the End of Man*. Cambridge, MA: Harvard University Press, 2020.

Beri, Emiliano. "Corallatori e guerra di corsa tra Sardegna e Corsica (1755–1768)." *Rives méditerranéennes* 57 (2018): 71–87.

Bernstein, Robin. *Racial Innocence: Performing American Childhood and Race from Slavery to Civil Rights*. New York: New York University Press, 2011.

Blackmun, Barbara W. "Iwebo and the White Men in Benin." In *Through African Eyes: The European in African Art, 1500 to Present*, edited by Nii O. Quarcoopome and Veit Arlt, 27–38. Detroit: Detroit Institute of Arts, 2009.

Bloom, Harold, and Jesse Zuba, eds. *American Religious Poems*. New York: Library of America, 2006.

Blum, Hester. "'Bitter with the Salt of Continents': Rachel Carson and Oceanic Returns." *Women's Studies Quarterly* 45, nos. 1–2 (Spring/Summer 2017): 287–91.

———. "The Prospect of Oceanic Studies." *PMLA/Publications of the Modern Language Association of America* 125, no. 3 (2010): 670–77.

Bowen, James. *The Coral Reef Era: From Discovery to Decline: A History of Scientific Investigation from 1600 to the Anthropocene Epoch*. New York: Springer, 2015.

Boydston, Jeanne. *Home and Work: Housework, Wages, and the Ideology of Labor in the Early Republic*. New York: Oxford University Press, 1990.

Braudel, Fernand. "History and the Social Sciences: The Longue Durée." 1958. In *The Longue Durée and World-Systems Analysis*, edited by Richard Edward Lee and Immanuel Wallerstein, 241–76. Albany: State University of New York Press, 2012.

Braverman, Irus. *Coral Whisperers: Scientists on the Brink*. Berkeley: University of California Press, 2018.

Brienen, Rebecca P. "Albert Eckhout's *African Woman and Child* (1641): Ethnographic Portraiture, Slavery, and the New World Subject." In *Slave Portraiture in the Atlantic World*, edited by Agnes Lugo-Ortiz and Angela Rosenthal, 229–55. New York: Cambridge University Press, 2013.

Brodhead, Richard. *Cultures of Letters: Scenes of Reading and Writing in Nineteenth-Century America*. Chicago: University of Chicago Press, 1993.

Brown, Bill. "The Origin of the American Work of Art." *American Literary History* 25, no. 4 (2013): 772–802.

———. *A Sense of Things: The Object Matter of American Literature*. Chicago: University of Chicago Press, 2010.

Brown, Laura. *Fables of Modernity: Literature and Culture in the English Eighteenth Century*. Ithaca, NY: Cornell University Press, 2001.

Browne, Janet. *Charles Darwin: A Biography*. Vol. 1, *Voyaging*. Princeton, NJ: Princeton University Press, 1995.

Browne, Neil W. *The World in Which We Occur: John Dewey, Pragmatist Ecology, and American Ecological Writing in the Twentieth Century*. Tuscaloosa: University of Alabama Press, 2007.

Buckridge, Steeve O. *The Language of Dress: Resistance and Accommodation in Jamaica, 1760–1890*. Kingston, Jamaica: University of West Indies Press, 2004.

Burnard, Trevor. *Jamaica in the Age of Revolution*. Philadelphia: University of Pennsylvania Press, 2020.

Buti, Gilbert. "Du rouge pour le Noir: Du corail méditerranéen pour la traite négrière au xviiie siècle." *Rives méditerranéennes* 57 (2018): 109–27.

Calcagno, Paolo. "A caccia dell'oro rosso: Le comunità del ponente ligure e la pesca del corallo nel xvii secolo." *Rives méditerranéennes* 57 (2018): 17–34.

Callisen, S. A. "The Evil Eye in Italian Art." *Art Bulletin* 19, no. 3 (September 1937): 450–62.

Calvert, Karin Lee Fishbeck. *Children in the House: The Material Culture of Early Childhood, 1600–1900*. Boston: Northeastern University Press, 1992.

Carp, E. Wayne. *Adoption in America: Historical Perspectives*. Ann Arbor: University of Michigan Press, 2009.

Chandler, Karen. "'Ye Are Builders': Child Readers in Frances Harper's Vision of an Inclusive Black Poetry." In *Who Writes for Black Children?*, edited by Katharine Capshaw and Anna Mae Duane, 41–58. Minneapolis: University of Minnesota Press, 2017.

Chow, Juliana. *Nineteenth-Century American Literature and the Discourse of Natural History*. New York: Cambridge University Press, 2021.

Chuh, Kandice. "It's Not about Anything." *Social Text* 32, no. 4 (2014): 125–34.

Cole, Michael. "Cellini's Blood." *Art Bulletin* 81, no. 2 (1999): 215–35.

Cooper, Helen A. *Winslow Homer Watercolors*. Washington, DC: National Gallery of Art, 1986.

Coriale, Danielle. "When Zoophytes Speak: Polyps and Naturalist Fantasy in the Age of Liberalism." *Nineteenth-Century Contexts* 34, no. 1 (2012): 19–36.

Crenshaw, Kimberlé Williams. "Race, Reform, and Retrenchment: Transformation and Legitimation in Antidiscrimination Law." *Harvard Law Review* 101, no. 7 (1988): 1331–87.

Curran, Andrew S. *The Anatomy of Blackness: Science and Slavery in an Age of Enlightenment*. Baltimore, MD: Johns Hopkins University Press, 2011.

Cutter, Martha J. "I Now Pronounce You Man and White: Racial Passing and Gender in Charles Chesnutt's Fiction." *American Literary Realism* 52, no. 3 (2020): 189–210.

Dain, Bruce R. *A Hideous Monster of the Mind: American Race Theory in the Early Republic*. Cambridge, MA: Harvard University Press, 2002.

Daston, Lorraine, and Fernando Vidal, eds. *The Moral Authority of Nature*. Chicago: University of Chicago Press, 2010.

Davidson, Kathleen. *Photography, Natural History and the Nineteenth-Century Museum: Exchanging Views of Empire*. New York: Routledge, 2017.

Davis, Angela Y. *Women, Race and Class*. New York: Random House, 1981.

Davis, Russell H. *Black Americans in Cleveland from George Peake to Carl B. Stokes, 1796–1969*. Washington, DC: Associated Publishers, 1985.

Dawson, Kevin. *Undercurrents of Power: Aquatic Culture in the African Diaspora*. Philadelphia: University of Pennsylvania Press, 2018.

Dawson, Virginia Parker. *Nature's Enigma: The Problem of the Polyp in the Letters of Bonnet, Trembley and Réaumur*. Philadelphia: American Philosophical Society, 1987.

Dayan, Colin. *The Law Is a White Dog: How Legal Rituals Make and Unmake Persons*. Princeton, NJ: Princeton University Press, 2013.

Delbourgo, James. *Collecting the World: Hans Sloane and the Origins of the British Museum*. Cambridge, MA: Harvard University Press, 2019.

Dillon, Elizabeth. *The Gender of Freedom: Fictions of Liberalism and the Literary Public Sphere*. Stanford, CA: Stanford University Press, 2004.

———. *New World Drama: The Performative Commons in the Atlantic World, 1649–1849*. Durham, NC: Duke University Press, 2014.

Dobbs, David. *Reef Madness: Charles Darwin, Alexander Agassiz, and the Meaning of Coral*. New York: Pantheon, 2009.

Doyle, Nora. *Maternal Bodies: Redefining Motherhood in Early America*. Chapel Hill: University of North Carolina Press, 2018.

Duggins, Molly. "Pacific Ocean Flowers: Colonial Seaweed Albums." In *The Sea and Nineteenth-Century Anglophone Literary Culture*, edited by Steve Mentz and Martha Elena Rojas, 119–34. New York: Taylor and Francis, 2016.

Elias, Ann. *Coral Empire: Underwater Oceans, Colonial Tropics, Visual Modernity*. Durham, NC: Duke University Press, 2019.

Elleray, Michelle. *Victorian Coral Islands of Empire, Mission, and the Boys' Adventure Novel*. New York: Routledge, 2019.

Elliott, John H. *Beware the Evil Eye: The Evil Eye in the Bible and the Ancient World*. Vol. 2, *Greece and Rome*. Eugene, OR: Cascade Books, 2016.

Ellis, Cristin. *Antebellum Posthuman: Race and Materiality in the Mid-Nineteenth Century*. New York: Fordham University Press, 2018.

Endt-Jones, Marion. "A Monstrous Transformation: Coral in Art and Culture." In *Coral: Something Rich and Strange*, edited by Marion Endt-Jones, 8–19. Liverpool: Liverpool University Press, 2013.

Entrikin, Isabelle Webb. *Sarah Josepha Hale and Godey's Lady's Book*. Lancaster, PA: Lancaster Press, 1946.

Fahs, Alice. *The Imagined Civil War: Popular Literature of the North & South, 1861–1865*. Chapel Hill: University of North Carolina Press, 2003.

Fielder, Brigitte. *Relative Races: Genealogies of Interracial Kinship in Nineteenth-Century America*. Durham, NC: Duke University Press, 2020.

Finley, Ruth Elbright. *The Lady of Godey's, Sarah Josepha Hale*. New York: J. B. Lippincott, 1931.

Fleissner, Jennifer. "Historicism Blues." *American Literary History* 25, no. 4 (2013): 699–717.

Foote, Stephanie. *Regional Fictions: Culture and Identity in Nineteenth-Century American Literature*. Madison: University of Wisconsin Press, 2001.

Foster, Frances Smith. *A Brighter Coming Day: A Frances Ellen Watkins Harper Reader*. New York: Feminist Press, 1990.

———. *Written by Herself: Literary Production by African American Women, 1746–1892*. Bloomington: Indiana University Press, 1993.

Fredrickson, George M. *The Black Image in the White Mind: The Debate on Afro-American Character and Destiny, 1817–1914*. New York: Harper and Row, 1971.

Frohne, Andrea E. *The African Burial Ground in New York City: Memory, Spirituality, and Space*. Syracuse, NY: Syracuse University Press, 2015.

Gardner, Eric. "Townsend, Jonas Holland." In *Oxford African American Studies Center*. Article published May 31, 2013. https://oxfordaasc.com/.

Gayles, LaMar. "Crafting and Forging the Black Diaspora: Black Histories in Jewelry." Master's thesis, Saint Olaf College, 2021.

Ghidiglia, Carlo. "L'industria del Corallo in Italia." *Giornale Degli Economisti* 5, no. 3 (1892): 479–510.

Gibson, Susannah. *Animal, Vegetable, Mineral?: How Eighteenth-Century Science Disrupted the Natural Order*. Oxford: Oxford University Press, 2015.

Gikandi, Simon. *Slavery and the Culture of Taste*. Princeton, NJ: Princeton University Press, 2011.

Giles, Paul. *Antipodean America: Australasia and the Constitution of US Literature*. New York: Oxford University Press, 2013.

Gillman, Susan. "Oceans of 'Longues Durées.'" *PMLA/Publications of the Modern Language Association of America* 127, no. 2 (2012): 328–34.

———. "The Political, the Personal, and 'The Function of American Literary Criticism at the Present Time,' 1983–2021." *American Literary History* 34, no. 1 (2022): 174–85.

———. "'Sure Identifiers': Race, Science, and the Law in *Pudd'nhead Wilson*." In *Mark Twain's "Pudd'nhead Wilson": Race, Conflict, and Culture*, edited by Susan Gillman and Forrest G. Robinson, 86–104. Durham, NC: Duke University Press, 1990.

Goddu, Theresa. *Selling Antislavery: Abolition and Mass Media in Antebellum America*. Philadelphia: University of Pennsylvania Press, 2020.

Gosetti-Ferencei, Jennifer. "On Literary Understanding." In *Varieties of Understanding: New Perspectives from Philosophy, Psychology, and Theology*, edited by Stephen R. Grimm, 67–92. Oxford: Oxford University Press, 2019.

Greeson, Jennifer Rae. *Our South: Geographic Fantasy and the Rise of National Literature*. Cambridge, MA: Harvard University Press, 2010.

Grosz, Elizabeth. *Time Travels: Feminism, Nature, Power*. Durham, NC: Duke University Press, 2005.

Guillaumin, Colette. *Racism, Sexism, Power and Ideology*. New York: Routledge, 1995.

Hancock, David. *Citizens of the World: London Merchants and the Integration of the British Atlantic Community, 1735–1785*. Cambridge: Cambridge University Press, 1995.

Handler, Jerome S. "The Middle Passage and the Material Culture of Captive Africans." *Slavery and Abolition* 30, no. 2 (2009): 1–26.

Hansen, Abbey. "Coral in Children's Portraits: A Charm against the Evil Eye." *Antiques* 120, no. 6 (December 1981): 1424–31.

Haraway, Donna J. *Simians, Cyborgs, and Women: The Reinvention of Nature*. New York: Routledge, 1991.

Harding, Jennifer Riddle. "Narrating the Family in Charles W. Chesnutt's 'Her Virginia Mammy.'" *Journal of Narrative Theory* 42, no. 3 (Fall 2012): 309–31.

Harris, Jennifer. "Barbara E. Pope (1854–1908)." *Legacy: A Journal of American Women Writers* 32, no. 2 (2015): 281–97.

Hart, Emma. *Trading Spaces: The Colonial Marketplace and the Foundations of American Capitalism*. Chicago: University of Chicago Press, 2020.

Hartman, Saidiya. *Lose Your Mother: A Journey along the Atlantic Slave Route*. New York: Farrar, Straus and Giroux, 2006.

Hazard, Sonia. "Evangelical Encounters: The American Tract Society and the Rituals of Print Distribution in Antebellum America." *Journal of the American Academy of Religion* 88, no. 1 (2020): 200–234.

Helmreich, Stefan, with contributions from Sophia Roosth and Michele Friedner. *Sounding the Limits of Life: Essays in the Anthropology of Biology and Beyond*. Princeton, NJ: Princeton University Press, 2015.

Herbert, Sandra. *Charles Darwin, Geologist*. Ithaca, NY: Cornell University Press, 2005.

Hood, Graham. *The Governor's Palace in Williamsburg: A Cultural Study*. Williamsburg, VA: Colonial Williamsburg Foundation, 1991.

Hood, Richard A. "Framing a 'Life in the Iron Mills.'" *Studies in American Fiction* 23, no. 1 (1995): 73–84.

Hughes, Heather A. "The Four Continents in Seventeenth-Century Embroidery." In *Personification: Embodying Meaning and Emotion*, edited by Walter S. Melion and Bart Ramakers, 716–49. Leiden: Brill, 2016.

Iannini, Christopher P. *Fatal Revolutions: Natural History, West Indian Slavery, and the Routes of American Literature*. Chapel Hill: University of North Carolina Press and Omohundro Institute, 2012.

Irmscher, Christoph. *Louis Agassiz: Creator of American Science*. New York: Houghton Mifflin Harcourt, 2013.

Isaacs, J. Susan. "Joshua Johnson." In *African American Lives*, edited by Henry Louis Gates Jr. and Evelyn Brooks Higginbotham, 459–60. New York: Oxford University Press, 2004.

Jackson, Cassandra. *Barriers between Us: Interracial Sex in Nineteenth-Century American Literature*. Bloomington: Indiana University Press, 2004.

Jenkins, Destin, and Justin Leroy, eds. *Histories of Racial Capitalism*. New York: Columbia University Press, 2021.

Johnson, Walter. *River of Dark Dreams: Slavery and Empire in the Cotton Kingdom*. Cambridge, MA: Harvard University Press, 2013.

Johnson, Wendy Dasler. *Antebellum American Woman's Poetry: A Rhetoric of Sentiment*. Carbondale: Southern Illinois University Press, 2016.

Jones, Jamie L. "Weathering, Wreck, Salvage: C19 Oceanic Time and the Metaphorical Turn in Oceanic Studies." Paper presented at *C19: The Society of Nineteenth-Century Americanists*, Coral Gables, FL, March 2022.

Jordan, Winthrop D. *White over Black: American Attitudes toward the Negro, 1550–1812*. Chapel Hill: University of North Carolina Press, 1968.

Kelley, Shannon. "The King's Coral Body: A Natural History of Coral and the Post-Tragic Ecology of *The Tempest*." *Journal for Early Modern Cultural Studies* 14, no. 1 (2014): 115–42.

Kendi, Ibram X. *Stamped from the Beginning: The Definitive History of Racist Ideas in America*. New York: Random House, 2017.

Kennedy-Nolle, Sharon D. *Writing Reconstruction: Race, Gender, and Citizenship in the Postwar South*. Chapel Hill: University of North Carolina Press, 2015.

Kimber, Marian Wilson. *The Elocutionists: Women, Music, and the Spoken Word*. Champaign: University of Illinois Press, 2017.

Kopytoff, Igor. "The Cultural Biography of Things: Commoditization as Process." In *The Social Life of Things: Commodities in Cultural Perspective*, edited by Arjun Appadurai, 65–91. Cambridge: Cambridge University Press, 1986.

LaCapra, Dominick. *Understanding Others: Peoples, Animals, Pasts*. Ithaca, NY: Cornell University Press, 2018.

LaFleur, Greta. *The Natural History of Sexuality in Early America*. Baltimore, MD: Johns Hopkins University Press, 2018.

Law, Robin. *The English in West Africa 1681–1683: The Local Correspondence of the Royal African Company of England, 1681–1699*. Part 1. Oxford: Oxford University Press, 1997.

Le Corbeiller, Clare. "Miss America and Her Sisters: Personifications of the Four Parts of the World." *Metropolitan Museum Art Bulletin* 19, no. 8 (1961): 209–23.

Lenhoff, Sylvia G., Howard M. Lenhoff, and Abraham Trembley. *Hydra and the Birth of Experimental Biology, 1744: Abraham Trembley's "Mémoires Concerning the Polyps."* Pacific Grove, CA: Boxwood Press, 1986.

"Literary Societies (Black)." In *Encyclopedia of Cleveland History*. Case Western Reserve University. Accessed April 15, 2022. https://case.edu/ech/articles/l/literary-societies-black.

Liverino, Basilio. *Red Coral: Jewel of the Sea*. Translated by Jane Helen Johnson. Bologna: Analisi, 1989.

Lo Basso, Luca, and Olivier Raveux. "Introduction. Le corail, un kaléidoscope pour l'étude de la Méditerranée dans le temps long." *Rives méditerranéennes* 57 (2018): 7–15.

Loeffelholz, Mary. *From School to Salon: Reading Nineteenth-Century American Women's Poetry*. Princeton, NJ: Princeton University Press, 2004.

Logan, Lisa M. "Race, Romanticism, and the Politics of Feminist Literary Study: Harriet Prescott Spofford's 'The Amber Gods.'" *Legacy: A Journal of American Women Writers* 18, no. 1 (2001): 35–51.

Long, Carolyn Morrow. *A New Orleans Voudou Priestess: The Legend and Reality of Marie Laveau*. Gainesville: University of Florida Press, 2005.

Lopez, Olivier. "Coral Fishermen in 'Barbary' in the Eighteenth Century: Between Norms and Practices." In *Law, Labour and Empire: Comparative Perspectives on Seafarers, c. 1500–1800*, edited by Maria Fusaro, Bernard Allaire, Richard J. Blakemore, and Tijl Vanneste, 195–211. New York: Palgrave Macmillan, 2015.

Lovell, Margaretta M. *Art in a Season of Revolution: Painters, Artisans, and Patrons in Early America*. Philadelphia: University of Pennsylvania Press, 2007.

Lustig, A. J. "Ants and the Nature of Nature in Auguste Forel, Erich Wasmann, and William Morton Wheeler." In *The Moral Authority of Nature*, edited by Lorraine Daston and Fernando Vidal, 282–307. Chicago: University of Chicago Press, 2010.

Martinez-Ruiz, Barbaro. "Sketches of Memory: Visual Encounters with Africa in Jamaican Culture." In *Art and Emancipation in Jamaica: Isaac Mendes Belisario and His Worlds*, edited by Tim Barringer, Gillian Forrester, and Barbaro Martinez-Ruiz, 103–17. New Haven, CT: Yale Center for British Art, 2007.

Mastroianni, Dominic. *Politics and Skepticism in Antebellum American Literature*. Cambridge: Cambridge University Press, 2014.

Mattingly, Carol. *Water Drops from Women Writers: A Temperance Reader*. Carbondale: Southern Illinois University Press, 2001.

McClinton, Katharine Morrison. *Antiques of American Childhood*. New York: Bramhall House, 1970.

McHenry, Elizabeth. *Forgotten Readers: Recovering the Lost History of African American Literary Societies*. Durham, NC: Duke University Press, 2002.

McIntyre, Rebecca C. "Promoting the Gothic South." *Southern Cultures* 11, no. 2 (2005): 33–61.

McKittrick, Katherine. *Dear Science and Other Stories*. Durham, NC: Duke University Press, 2020.

Meer, Sarah. *Uncle Tom Mania: Slavery, Minstrelsy, and Transatlantic Culture in the 1850s*. Athens: University of Georgia Press, 2005.

Merish, Lori. *Archives of Labor: Working-Class Women and Literary Culture in the Antebellum United States*. Durham, NC: Duke University Press, 2017.

———. *Sentimental Materialism: Gender, Commodity Culture, and Nineteenth-Century American Literature*. Durham, NC: Duke University Press, 2000.

Meyers, Amy R. W. "Picturing a World in Flux: Mark Catesby's Response to Environmental Interchange and Colonial Expansion." In *Empire's Nature: Mark Catesby's New World Vision*, edited by Amy R. W. Meyers and Margaret Beck Pritchard, 228–62. Chapel Hill: University of North Carolina Press and Omohundro Institute, 2012.

Moody-Turner, Shirley. "Folklore and African American Literature in the Post-Reconstruction Era." In *A Companion to African American Literature*, edited by Gene Andrew Jarrett, 200–211. Hoboken, NJ: Blackwell, 2010.

Mott, Frank Luther. *A History of American Magazines, 1850–1865*. Vol. 2. Cambridge, MA: Harvard University Press, 1938.

Murison, Justine S. *The Politics of Anxiety in Nineteenth-Century American Literature*. New York: Cambridge University Press, 2011.

Narain, Denise deCaires. *Contemporary Caribbean Women's Poetry: Making Style*. New York: Routledge, 2003.

Navakas, Michele, and Dominic Mastroianni. "Emerson Undersea." Chap. 4 in *The Oxford Handbook of Ralph Waldo Emerson*, edited by Christopher Hanlon. New York: Oxford University Press, forthcoming.

Neely, Michelle C. *Against Sustainability: Reading Nineteenth-Century America in the Age of Climate Crisis*. New York: Fordham University Press, 2020.

Nelson, Dana D. *Commons Democracy: Reading the Politics of Participation in the Early United States*. New York: Fordham University Press, 2016.

Neumann, Birgit, and Jan Rupp. "Sea Passages: Cultural Flows in Caribbean Poetry." *Atlantic Studies* 13, no. 4 (2016): 472–90.

Neumann, Brian C. "Black Virginians in Blue: Black Union Soldiers and Sailors from Albemarle County, Virginia." *Journal of Slavery and Data Preservation* 2, no. 2 (2021): 12–18.

———. "Lee, Frank." In *Oxford African American Studies Center*. Article published January 28, 2022. https://oxfordaasc.com.

Nozomu, Iwasaki, ed. *Precious Coral and the Legacy of the Coral Road*. Newcastle-upon-Tyne: Cambridge Scholars Publisher, 2021.

Ogundiran, Akinwumi. "Of Small Things Remembered: Beads, Cowries, and Cultural Translations of the Atlantic Experience in Yorubaland." *International Journal of African Historical Studies* 35, nos. 2/3 (2002): 427–57.

Ogundiran, Akinwumi, and Paula Saunders, eds. *Materialities of Ritual in the Black Atlantic*. Bloomington: Indiana University Press, 2014.

Olney, James. "'I Was Born': Slave Narratives, Their Status as Autobiography and as Literature." *Callaloo*, no. 20 (1984): 46–73.

Orlowski-Yang, Jeff, dir. *Chasing Coral*. Exposure Labs, 2017. 1 hr., 29 min. https://www.chasingcoral.com/.

Otter, Samuel. "Stowe and Race." In *The Cambridge Companion to Harriet Beecher Stowe*, edited by Cindy Weinstein, 15–38. Cambridge: Cambridge University Press, 2004.

Parrish, Susan Scott. *American Curiosity: Cultures of Natural History in the Colonial British Atlantic World*. Chapel Hill: University of North Carolina Press and Omohundro Institute, 2006.

———. "Rummaging / In and Out of Holds." *Early American Literature* 45, no. 2 (2010): 261–74. http://www.jstor.org/stable/27856619.

Patterson, Orlando. *Slavery and Social Death: A Comparative Study*. Cambridge, MA: Harvard University Press, 1982.

Pears, Sean. "A Speculative Reading of Black Feminist Resistance in George Washington Cable's *The Grandissimes*." *Arizona Quarterly: A Journal of American Literature, Culture, and Theory* 76, no. 2 (2020): 115–40.

Peterson, Carla. *"Doers of the Word": African-American Women Speakers and Writers in the North, 1830–80*. New York: Oxford University Press, 1995.

Peterson, Norma L. *Freedom and Franchise: The Political Career of B. Gratz Brown*. Columbia: University of Missouri Press, 1965.

Pettigrew, Andrew. *Freedom's Debt: The Royal African Company and the Politics of the Atlantic Slave Trade, 1672–1752*. Chapel Hill: University of North Carolina Press, 2013.

Pfaelzer, Jean. *Parlor Radical: Rebecca Harding Davis and the Origins of American Social Realism.* Pittsburgh: University of Pittsburgh Press, 1996.

Philbrick, Nathaniel. *Sea of Glory: America's Voyage of Discovery, the US Exploring Expedition, 1838–1842.* New York: Penguin, 2004.

Polcha, Elizabeth. "Voyeur in the Torrid Zone: John Gabriel Stedman's *Narrative of a Five Years Expedition against the Revolted Negroes of Surinam, 1773–1838.*" *Early American Literature* 54, no. 3 (2019): 673–710.

Porter, Carolyn. "Roxana's Plot." In *Mark Twain's "Pudd'nhead Wilson": Race, Conflict, and Culture,* edited by Susan Gillman and Forrest G. Robinson, 121–36. Durham, NC: Duke University Press, 1990.

Putzi, Jennifer. *Fair Copy: Relational Poetics and Antebellum American Women's Poetry.* Philadelphia: University of Pennsylvania Press, 2021.

Raimon, Eve Allegra. *The "Tragic Mulatta" Revisited: Race and Nationalism in Nineteenth-Century Antislavery Fiction.* New Brunswick, NJ: Rutgers University Press, 2004.

Ranta, Judith A. *Women and Children of the Mills: An Annotated Guide to Nineteenth-Century American Textile Factory Literature.* Westport, CT: Greenwood Press, 1999.

Reveal, James L. "Identification of the Plants and Animals Illustrated by Mark Catesby for His *Natural History of Carolina, Florida, and the Bahama Islands.*" *Phytoneuron: Digital Publications in Plant Biology* 6 (2013): 1–55. https://www.phytoneuron.net/2013-publications/.

Riss, Arthur. *Race, Slavery, and Liberalism in Nineteenth-Century American Literature.* Cambridge: Cambridge University Press, 2006.

Roach, Joseph. *Cities of the Dead: Circum-Atlantic Performance.* New York: Columbia University Press, 1996.

Robinson, Greg. "Cleveland." In *Encyclopedia of African-American Culture and History,* edited by Jack Salzman, David L. Smith, and Cornel West, 2:595–99. New York: Macmillan Library Reference, 1996.

Rockman, Seth. *Scraping By: Wage Labor, Slavery, and Survival in Early Baltimore.* Baltimore, MD: Johns Hopkins University Press, 2009.

Roediger, David. *The Wages of Whiteness: Race and the Making of the American Working Class.* New York: Verso, 1991.

Roger, Jacques. *The Life Sciences in Eighteenth-Century French Thought.* Stanford, CA: Stanford University Press, 1997.

Rogers, Dorothy G. *America's First Women Philosophers: Translating Hegel, 1860–1925.* New York: Bloomsbury, 2005.

Ross, Felicia G. Jones. "Fragile Equality: A Black Paper's Portrayal of Race Relations in Late 19th-Century Cleveland." *Howard Journal of Communications* 6, nos. 1–2 (1995): 53–68. https://doi.org/10.1080/10646179509361684.

Rossi, Sergio. *Oceans in Decline.* New York: Springer, 2019.

Rudwick, Martin J. S. *Worlds before Adam: The Reconstruction of Geohistory in the Age of Reform.* Chicago: University of Chicago Press, 2008.

Rusert, Britt. *Fugitive Science: Empiricism and Freedom in Early African American Culture.* New York: New York University Press, 2017.

Ryder, A. F. C. *Benin and the Europeans, 1485–1897.* London: Longmans, 1969.

Sánchez-Eppler, Karen. *Touching Liberty: Abolition, Feminism, and the Politics of the Body.* Berkeley: University of California Press, 1993.

Schakenbach Regele, Lindsay. "A Brief History of the History of Capitalism, and a New American Variety." Unpublished manuscript. Last modified 2022.

———. *Manufacturing Advantage: War, the State, and the Origins of American Industry, 1776–1848.* Baltimore, MD: Johns Hopkins University Press, 2019.

Sharpe, Christina. *In the Wake: On Blackness and Being.* Durham, NC: Duke University Press, 2016.

Sherrard-Johnson, Cherene. "Sensing Black Coral." *ISLE: Interdisciplinary Studies in Literature and Environment* (July 2021), https://doi.org/10.1093/isle/isab050.

Silva, Cristobal. "The Silent History of Nostalgia." Unpublished manuscript. Last modified 2022.

Singh, Nikhil Pal. *Race and America's Long War*. Berkeley: University of California Press, 2017.

Sitch, Bryan, and Keith Sugden. "Perseus, Medusa, and the Birth of Coral." In *Coral: Something Rich and Strange*, edited by Marion Endt-Jones, 41–44. Liverpool: Liverpool University Press, 2013.

Sivasundaram, Sujit. *Nature and the Godly Empire: Science and Evangelical Mission in the Pacific, 1795–1850*. New York: Cambridge University Press, 2005.

Smedman, M. Sarah. "Sarah Josepha (Buell) Hale." In *American Writers for Children before 1900*, ed. Glenn E. Estes, 207–17. Vol. 42 of *Dictionary of Literary Biography*. Detroit: Gale Literature Resource Center, 1985.

Smith, Linda Tuhiwai. *Decolonizing Methodologies: Research and Indigenous Peoples*. London: Zed Books, 1999.

Sonstegard, Adam. "The Graphic African, the Illustrator's Gaze, and *The Grandissimes*." *Mississippi Quarterly* 72, no. 2 (2019): 195–232.

Spatz, David A. "Barnett, Ferdinand L." In *Oxford African American Studies Center*. Article published December 1, 2009. https://oxfordaasc.com/.

Spicer, Joneath. "The Personification of Africa with an Elephant-Head Crest in Cesare Ripa's *Iconologia* (1603)." In *Personification: Embodying Meaning and Emotion*, edited by Walter S. Melion and Bart Ramakers, 677–715. Leiden: Brill, 2016.

Spillers, Hortense J. *Black, White, and in Color: Essays on American Literature and Culture*. Chicago: University of Chicago Press, 2003.

Spires, Derrick. "Genealogies of Black Modernities." *American Literary History* 32, no. 4 (2020): 611–22.

———. *The Practice of Citizenship: Black Politics and Print Culture in the Early United States*. Philadelphia: University of Pennsylvania Press, 2019.

Sponsel, Alistair. *Darwin's Evolving Identity: Adventure, Ambition, and the Sin of Speculation*. Chicago: University of Chicago Press, 2018.

St. John A.M.E. Church (Cleveland, OH). *150th Anniversary, St. John A.M.E. Church: Souvenir Journal, 1830–1980, 150th Year*. Cleveland: St. John A.M.E. Church, 1980.

"St. John's African Methodist Episcopal (AME) Church." *Encyclopedia of Cleveland History*. Case Western Reserve University. Accessed April 15, 2022. https://case.edu/ech/articles/s/st-johns-african-methodist-episcopal-ame-church.

Stepan, Nancy. *The Idea of Race in Science: Great Britain, 1800–1960*. London: Palgrave Macmillan Limited, 1982.

Strang, Cameron B. *Frontiers of Science: Imperialism and Natural Knowledge in the Gulf South Borderlands, 1500–1850*. Chapel Hill: University of North Carolina Press and Omohundro Institute, 2018.

Sundquist, Eric J. *To Wake the Nations: Race in the Making of American Literature*. Cambridge, MA: Harvard University Press, 1998.

Tamarkin, Elisa. "Reading for Relevance." *Commonplace: The Journal of Early American Life* 10, no. 4 (July 2010), http://commonplace.online/article/reading-for-relevance/.

Thorndike, Lynn. *A History of Magic and Experimental Science*. Vol. 8, *The Seventeenth Century*. New York: Columbia University Press, 1923.

Tichi, Cecelia. "Introduction: Cultural and Historical Background." In *Life in the Iron-Mills*, edited by Cecelia Tichi, 3–25. Boston: Bedford Cultural Editions, 1998.

Torntore, Susan J. "Precious Red Corals: Markets and Meanings." *Beads: Journal of the Society of Bead Researchers* 16 (2004): 3–16.

Trivellato, Francesca. *The Familiarity of Strangers: The Sephardic Diaspora, Livorno, and Cross-Cultural Trade in the Early Modern Period*. New Haven, CT: Yale University Press, 2009.

Trouillot, Michel-Rolph. *Silencing the Past: Power and the Production of History*. Boston: Beacon Press, 1995.

Tuennerman-Kaplan, Laura. *Helping Others, Helping Ourselves: Power, Giving, and Community Identity in Cleveland, Ohio, 1880–1930*. Kent, OH: Kent State University Press, 2011.

Turner, Arlin. *George W. Cable: A Biography*. Baton Rouge: Louisiana State University Press, 1966.

Vandersmissen, Jan. "Experiments and Evolving Frameworks of Scientific Exploration: Jean-André Peyssonnel's Work on Coral." In *Expeditions as Experiments: Practising Observation and Documentation*, edited by Marianne Klemun and Ulrike Spring, 51–72. New York: Springer, 2016.

Vermeren, Hugh. "Être corailleur en Algérie au xixe siècle: pratiques du métier et reconversion profes-
sionnelle chez une population maritime en déclin à l'époque coloniale (Bône, La Calle, 1832–1888)."
*Rives méditerranéennes* 57 (2018): 35–54.

von Kemnitz, Eva-Maria. "Porous Frontiers of the Hand Symbol." In *In the Iberian Peninsula and Be-
yond: A History of Jews and Muslims (15th–17th Centuries)*, edited by José Alberto, R. Silva Tavim,
Maria Filomena Lopes de Barros, and Lúcia Liba Mucznik, 2: 192–207. Newcastle-upon-Tyne, UK:
Cambridge Scholars Publishing, 2015.

Walls, Laura Dassow. *The Passage to Cosmos: Alexander von Humboldt and the Shaping of America.*
Chicago: University of Chicago Press, 2009.

———. *Seeing New Worlds: Henry David Thoreau and Nineteenth-Century Natural Science.* Madison:
University of Wisconsin Press, 1995.

Ward, Martha. *Voodoo Queen: The Spirited Lives of Marie Laveau.* Jackson: University Press of Mississippi, 2004.

Wardley, Lynn. "Relic, Fetish, Femmage: The Aesthetics of Sentiment in the Work of Stowe." *Yale Journal
of Criticism* 5, no. 3 (1992): 165–91.

Warsh, Molly. *American Baroque: Pearls and the Nature of Empire.* Chapel Hill: University of North
Carolina Press and Omohundro Institute, 2018.

Waterman, Bryan. *Republic of Intellect: The Friendly Club of New York City and the Making of American
Literature.* Baltimore, MD: Johns Hopkins University Press, 2007.

Weekley, Carolyn J., and Stiles Tuttle Colwill. With Leroy Graham and Mary Ellen Hayward. *Joshua
Johnson: Freeman and Early American Portrait Painter.* Baltimore, MD: Abby Aldrich Rockefeller
Folk Art Center, Colonial Williamsburg Foundation / Maryland Historical Society, 1987.

Wees, Beth Carver, and Medill Higgins Harvey. *Early American Silver in the Metropolitan Museum of Art.*
New York: Metropolitan Museum of Art, 2013.

Weheliye, Alexander G. *Habeas Viscus: Racializing Assemblages, Biopolitics, and Black Feminist Theories
of the Human.* Durham, NC: Duke University Press, 2014.

Weiss, Harry Bischoff. *American Baby Rattles from Colonial Times to the Present.* Trenton, NJ: private
printer, 1941.

Welsh, Stephen Terence. "Red Coral Beads, Benin City (nineteenth century)." In *Coral: Something Rich
and Strange*, edited by Marion Endt-Jones, 49–52. Liverpool: Liverpool University Press, 2013.

Wheeler, Edmund. *The History of Newport, New Hampshire.* Concord, NH: Republican Press Association, 1879.

Wheeler, Roxann. *The Complexion of Race: Categories of Difference in Eighteenth-Century British Culture.*
Philadelphia: University of Pennsylvania Press, 2000.

Wiegman, Robyn. *American Anatomies: Theorizing Race and Gender.* Durham, NC: Duke University
Press, 1995.

Wood, Peter H., and Karen C. C. Dalton. *Winslow Homer's Images of Blacks: The Civil War and Reconstruction Years.* Austin: University of Texas Press, 1988.

Wright, Nazera Sadiq. *Black Girlhood in the Nineteenth Century.* Champaign: University of Illinois Press,
2016. ProQuest EBook.

Yentsch, Anne. "Beads as Silent Witnesses of an African-American Past: Social Identity and the Artifacts of Slavery in Annapolis, Maryland." *Kroeber Anthropological Society Papers* 79 (1995): 44–60.

Yogev, Gedalia. *Coral and Diamonds: Anglo-Dutch Jews and Eighteenth-Century Trade.* Leicester: Leicester University Press, 1978.

Young, Hilary. *English Porcelain, 1745–95: Its Makers, Design, Marketing and Consumption.* London:
V and A Publications, 1999.

# Index